Discovering Mathematics

A Guide for Curriculum Leaders and Administrators

Project Editor: Joan Lewis

Editorial Assistants: Brian Collins, Aneesa Davenport

Series Authors: Michael Serra, Jerald Murdock,
Ellen Kamischke, Eric Kamischke

Reviewer: Steven Leinwand

Consulting Writer: Larry Wiley

Production Editor: Christine Osborne

Copyeditor: Deborah Cogan

Production Director: Diana Jean Ray

Production Coordinator and Text Designer: Charice Silverman

Compositor: The Cowans

Art Editor: Jason Luz

Photo Researcher: Margee Robinson

Art and Design Coordinator: Kavitha Becker

Cover Designer: Todd Bushman

Cover Photos: Ken Karp Photography

Credits: Ken Karp Photography, pages 4, 9, 12, 20, 23

Prepress and Printer: Data Reproductions

National Sales Director: Kelvin Taylor

Executive Editor: Casey FitzSimons

Publisher: Steven Rasmussen

Key Curriculum Press
1150 65th Street
Emeryville, CA 94608
510-595-7000
editorial@keypress.com
sales@keypress.com
www.keypress.com

Printed in the United States of America

10 9 8 7 6 5 4 3 2   09 08 07 06 05   ISBN 1-55953-662-4

Contents

Executive Summary: A Short Overview of the *Discovering Mathematics* Series

The three courses of the *Discovering Mathematics* series encompass algebra, geometry, and advanced algebra through engaging, discovery-based investigations and are highly consistent with state and national standards-based efforts to improve school mathematics. These courses can very well meet the needs of your teachers and their students for high-quality, easy-to-use mathematics materials.

This series is proven effective. It will help you enhance and reshape the mathematics program in your school. The effectiveness of the *Discovering Mathematics* approach has been verified by research and investigations conducted with students who have been using this series. It is clear that these students have learned and performed better on assessments than comparable non-users did.

Each course strongly emphasizes active learning through a rich variety of strategies and activities for teaching, learning, and assessment. Students following this curriculum will advance academically and will be prepared for critical assessments and testing.

Technology—including very significant use of graphing calculators and software such as The Geometer's Sketchpad® dynamic geometry and Fathom Dynamic Statistics™, in addition to the Web—is regularly and routinely used as an integral component of teaching and learning.

The series provides critically important ideas and support for its successful implementation, especially for ongoing professional development of teachers. Many kinds of support are available with this program, such as comprehensive Web sites for teachers, students, and parents, along with traditional printed resources.

Finally, of course, it is *crucial* to engage key parents and key teachers actively and constructively in the process of evaluating and recommending the adoption of new materials so that you enlist the enthusiastic support and endorsement of

- parents and parent groups (*especially* the parent-teacher organization) who can influence public and parent opinion in favor of a new curriculum. (See the section Building Parent Support for many useful ideas along these lines.)
- teachers and teacher groups (*especially* the teachers' union) who must buy in strongly to the process of adopting and fully implementing new curriculum. (See the section Supporting Teachers for many useful ideas along these lines.)

This series merits your serious consideration. Further information is available from Kelvin Taylor, National Sales Director, ktaylor@keypress.com, and Bettye Forte, National Sales Consultant, bforte@keypress.com. You can also obtain information by contacting sales@keypress.com, calling 800.995.6284, or visiting www.keypress.com.

Introduction: The *Discovering Mathematics* Approach

Helping students reach high levels of understanding and mastery of mathematics, helping them do well on tests, and giving them the skills employers value are ongoing challenges for teachers, curriculum leaders, and administrators. The *Discovering Mathematics* series from Key Curriculum Press is designed to bring about that mathematical learning.

Three textbooks comprise the *Discovering Mathematics* series: *Discovering Algebra: An Investigative Approach; Discovering Geometry: An Investigative Approach; Discovering Advanced Algebra: An Investigative Approach*

Discovering Mathematics is a discovery-based, investigation-based series designed for the three-year sequence of algebra, geometry, and advanced algebra courses offered in most schools. The student books in this series are appealing, full-color presentations of important mathematics supported by hands-on activities, easy-to-read text, examples, exercises, and enriching features. The illustrations include representations of a wide variety of students and real-world situations. The spacious book design enables easy teacher and student navigation through the chapters. Integrated text and art displays help convey complex mathematical concepts and processes. Each book covers traditional topics, includes up-to-date mathematical ideas, and uses modern pedagogy: multiple representations of functions, alternative solution methods, and problem-solving techniques. The series has been carefully designed to accommodate students with different learning styles and different ability levels.

Appropriate use of technology, including graphing calculators and computer software, keeps the emphasis on ideas. Students visualize and explore concepts by working with real-world data. The series enables them to streamline processes without sacrificing their skill development.

Each hard-bound student edition in the series has a companion hard-bound teacher's edition with reduced student pages and wrap-around commentary. A full set of printed and electronic teaching resources is provided; these materials include assessment and practice resources and technology supplements. Each student book is available electronically. Also, comprehensive companion Web sites, one for students and parents and another for teachers, have been provided as collateral support for each student book.

"[With *Discovering Algebra*] the classroom dynamic has changed. . . . Our teachers now see more 'light bulb' moments as their students make sense of the mathematics they are learning."

—Carol Treglio, Secondary Mathematics Resource Teacher, San Diego Unified School District

This guide explains the *Discovering Mathematics* approach in more detail, and then describes the extensive development and testing process each *Discovering Mathematics* book has undergone. (See the section What Makes the *Discovering Mathematics* Series So Effective? and Appendix A: The *Discovering Mathematics* Series Authors.) The guide elucidates the research supporting the *Discovering Mathematics* approach and provides information on assessing student achievement. (See the section What Is the Scientific Research Base for the *Discovering Mathematics* Series?)

This guide also offers helpful advice about what a teacher needs to learn in order to implement the series effectively. For example, see the Supporting Teachers section under the heading How Can the Implementation of the *Discovering Mathematics*

Series Be Effectively Supported? Helpful information for teachers can also be found in *Discovering Mathematics: A Guide for Teachers.* Additional information on professional development can be found under Pedagogy in the section What Makes the *Discovering Mathematics* Series So Effective?

Solid parental support is a crucial factor in successfully implementing the *Discovering Mathematics* series, especially if many parents expect to see a mathematics curriculum that is essentially unchanged from the one they encountered as students. You should seek out parents who can be opinion shapers and enlist their aid in facilitating the shift to better curriculum, instruction, and assessment. In the section on effective implementation, we provide a helpful discussion of this issue and references to the Web sites we have provided to help address the concerns of parents.

The appendices introduce the series authors, list the tables of contents of the books, describe specific features of the student editions and the teacher's editions, summarize the mathematical topics, and list resources on learning research.

Basic Facts about *Discovering Mathematics*

The following description of the *Discovering Mathematics* series explains the content and pedagogy, showing how it is similar to a traditional series and how it is unique.

- The books of the *Discovering Mathematics* series cover the mathematics that students should learn in algebra, geometry, and advanced algebra. They provide a thorough treatment of mathematical topics that prepare students for advanced mathematics courses as well as for technical careers. The series is strongly aligned with the National Council of Teachers of Mathematics *Principles and Standards for School Mathematics.* The series has also been correlated to most state standards and frameworks. These correlations are available from Key Curriculum Press.
- The *Discovering Mathematics* series uses a student-friendly, investigation-based or "discovery" learning method. Students' first exposure to a topic comes in the form of hands-on activities that they do in small groups facilitated by their teacher. They pool their thinking in class discussions and learn the correct vocabulary, conventions, and operations needed for the task at hand. They work through examples and do both pure-mathematics exercises and real-world applications, drawing diagrams and graphs and explaining their thinking.
- *Discovering Mathematics* embodies the best features of "constructivist" learning and teaching strategies in the investigations and follows these with concrete examples, explanations, and highlighted definitions. The mathematics is very evident in the books, which include skill-based and application exercises, boxed conjectures and definitions, and answers to selected problems.
- *Discovering Mathematics* differs from a traditional, directive approach that typically consists of the teacher presenting a lecture, working a few example exercises, and getting students started on doing more exercises of the same sort for their homework. In contrast, the *Discovering Mathematics* series uses hands-on investigations to enable students to discover the implications of patterns and processes before the mathematics is fully set out. The program develops student initiative, confidence, understanding, and the ability to communicate. Students get used to thinking through new problems without being overly dependent on the teacher telling them what to do and how to do it.
- Teachers who already allow students to work in cooperative groups (even if only on projects or to get started on homework) and who encourage and facilitate class discussions even once in a while will have the least amount of adjustment to make in their teaching styles in order to use *Discovering Mathematics.* Teachers who tend to read from their textbooks or who rely on a lecture approach will have the greatest adjustment to make. However, even those teachers probably ask questions and wait for students to answer, thereby encouraging at least some responsibility among students for their own learning. The development of student responsibility and self-reliance is at the heart of the investigative approach.
- The *Discovering Mathematics* series integrates technology, especially graphing calculators, so you'll need to ensure that these tools are readily available. *Discovering Algebra* assumes that all students will have easy access to graphing calculators during class time. Exercises that require calculators are clearly indicated so that teachers can assign them during class. *Discovering Advanced Algebra* assumes that all students will have graphing calculators in class and also at home. *Discovering Geometry* doesn't depend on students having graphing calculators on a regular basis, but there are a few clearly marked activities that do require them. Moreover, using dynamic geometry software

such as The Geometer's Sketchpad can yield great benefits (as can using other appropriate software). There are frequent opportunities to use Sketchpad™, especially in *Discovering Geometry,* whether the teacher does in-class demonstrations with a single, large-screen monitor or computer projector, students share a few classroom computers, or the class goes to a computer lab where each student can complete an exploration. All three books contain some optional activities that call for using The Geometer's Sketchpad or Fathom Dynamic Statistics.

A Short History of the *Discovering Mathematics* Series

For 30 years, Key Curriculum Press has been creating and publishing materials that facilitate mathematics learning for students and instruction for teachers. Originally concentrating on supplemental materials, manipulatives, and self-paced curriculum, Key received frequent inquiries about the availability of core curriculum textbooks from the enthusiastic users of our innovative materials.

In response, we published the first edition of *Discovering Geometry: An Inductive Approach* by Michael Serra in 1989. Its success and popularity among both students and teachers gave impetus to the demand for more—for a complete series of textbooks to address the sequence of beginning algebra–geometry–advanced algebra. The series would use the investigative approach pioneered in *Discovering Geometry,* an approach that in only a few years had been proven successful in mathematics classrooms and is now widely imitated.

In 1997, the second edition of *Discovering Geometry* delighted its loyal fans and new users with full-color art and design and a refined presentation. With the 1998 publication of *Advanced Algebra Through Data Exploration: A Graphing Calculator Approach* by Jerald Murdock, Eric Kamischke, and Ellen Kamischke, Key went further with cutting-edge graphing calculator and computer technology. Our goal was to help teachers who were eager to develop enlightened instructional strategies suitable for students living in a world full of new technologies.

At the same time, Key was already planning *Discovering Algebra: An Investigative Approach* with the same group of authors. The draft manuscript for *Discovering Algebra* was field-tested and then published as a Preliminary Edition, and was further revised before it appeared in 2002 as a full-color textbook complete with *Teacher's Edition* and a full package of support materials. With the publication of *Discovering Algebra,* Key was ready to offer a book for each course in the typical three-year, high school curriculum. We saw the need, though, to update and unify the look and the instructional elements of the preceding books in the series, to update the technology for forward-aiming classrooms, and to modernize social content and applications. Both the design and the pedagogy make the content suitable for students of diverse abilities, motivations, and wide-ranging educational destinations.

In 2003, *Discovering Geometry* was published in a third edition and, by 2004, *Advanced Algebra Through Data Exploration* was extensively revised to become *Discovering Advanced Algebra: An Investigative Approach.* This huge effort, a 15-year process, has resulted in the *Discovering Mathematics* series.

What Makes the *Discovering Mathematics* Series So Effective?

The books of the *Discovering Mathematics* series not only meet but exceed the expectations of the National Council of Teachers of Mathematics (NCTM) *Principles and Standards for School Mathematics* (PSSM), both in content and in process. Consequently, students learn mathematical concepts as they solve problems, reason (explain why and create rigorous proofs), communicate about their conjectures and solutions, and make connections to their previous mathematics knowledge and other disciplines. They also see and learn to appreciate multiple representations and solution methods. The *Discovering Mathematics* authors and publisher have given exhaustive attention to content and pedagogy, the intelligent use of technology, and students' development of problem solving and other life skills. The publisher provides many useful print, electronic, and Web-based resources for teachers, students, and parents.

"*Discovering Geometry*'s approach to learning and teaching geometry is the only way to teach geometry! The investigative and discovery approach with the use of inductive and deductive reasoning is a perfect combination in helping students enjoy the learning of geometry."

—Judy Hicks, Ralston Valley High School, Arvada, CO

Content

Discovering Mathematics thoroughly covers the traditional content of algebra, geometry, and advanced algebra courses. (See Appendix B for a table of contents for each book.) In *Discovering Algebra,* students learn about rates, linear relationships, and exponential and quadratic functions, as well as recursion, proportional reasoning, and exploration of data. In *Discovering Geometry,* they begin by studying the properties of geometric figures and inductive reasoning. Then, they discover the properties of polygons and circles, explore transformations, apply trigonometry, and build deductive systems. In *Discovering Advanced Algebra,* students move on to study functions and transformations; exponential, power, logarithmic, and polynomial functions; matrices and linear systems; and parametric equations and trigonometry.

Although the student books are designed for discrete courses in algebra, geometry, and advanced algebra, the algebra courses incorporate geometry and the geometry course incorporates algebra to build connections between the mathematical disciplines and to keep students' skills sharp.

"This approach to learning is the way of the future for our students. Our goal is to have 100% of our students succeed in mathematics. I love the *Discovering Algebra* and *Discovering Geometry* program. I'm a believer!"

—Margie Hill, District Coordinating Teacher, Blue Valley Schools, Overland Park, KS

Pedagogy

The *Discovering Mathematics* series focuses on helping all students build understanding. Teachers use hands-on investigations of interesting problems to engage students in cooperative groups using graphing calculators and sometimes computer software.

"This is my first year teaching with *Discovering Geometry.* By the end of the first quarter, my students were sold on this text and defended the value of cooperative learning against all comers. By the end of the year, I had one student go from a C to an A, primarily because she became so engaged with the cooperative learning process that she learned the material through sheer enthusiasm. While her other teachers were complaining about her lack of effort, I got to rave about how motivated she was! When the students' binders were all graded, they proposed we go around the circle and each tell what had been the best part of the class for them. I can't imagine ever wanting to do that when I was in high school! I'll be experimenting, learning, and fine-tuning for a long time to come, but what a great way to start!"

—Marion Athearn, Lincoln School, Providence, RI

In a hands-on investigation, students gather data and explore it to discover relationships, make conjectures, and solve problems. The teacher, acting as a lead investigator, facilitates the students' sharing of their results to bring out the mathematical ideas. As students share results, they see different problem-solving approaches and deepen their own learning by communicating about their thinking. *Discovering Mathematics* students see many ways to make the connections that lead to understanding. The teacher helps students to articulate their ideas with mathematical vocabulary, develop a common knowledge base, and observe mathematical notation and conventions.

"My students and I love the investigations in *Discovering Algebra.* The students pay more attention when activities are hands-on; they can interact as they think about what they're learning, and this helps them remember the math. Working in groups, they find they can depend on each other and are less dependent on the teacher. The discovery format definitely helps them to be more successful."

—Kerry Connelly, Blue Valley Northwest High School, Overland Park, KS

The *Discovering Mathematics* series supports teachers' efforts to facilitate cooperative-group work and enhance mathematics learning and understanding for several good reasons.

- Many students participate more actively in a small group than in a full class.
- Students who have not been successful in mathematics but who are comfortable in social situations can gain confidence in their mathematical abilities.
- As students articulate an idea for other group members, they deepen their understanding of it. They connect new ideas to their own experiences, thereby developing habits of mind that support making sense of new ideas and remembering them.
- Students are exposed to more ideas for solving problems—when solving challenging problems, "two heads are better than one." They learn to recognize the validity of alternative approaches, and thinking about as well as using a wide range of ways to solve problems enables students to apply the underlying mathematical concepts in new situations.
- Students learn to solve more complex problems than they could without a group to contribute different kinds of ideas and expertise.
- Students learn to respect and appreciate differences in learning styles, physical and mental abilities, and cultural or ethnic backgrounds.

Technology

Technology, including graphing calculators and computer software, is integrated throughout the *Discovering Mathematics* series, but it is used in a way that does not impede the learning of calculation skills. In fact, many exercises require calculation practice without technology. Students use technology primarily in the investigations to explore real data and visualize patterns and relationships.

In *Discovering Algebra* and *Discovering Advanced Algebra,* the primary technology tool is the graphing calculator. With it, students see how a graph changes as parts of its equation change—an investigation that would be tedious with pencil and paper. With a graphing calculator and, in some cases, also a motion sensor, students can gather and analyze "messy" real data. They can also apply algebra to analyze larger, previously collected and tabulated sets of real-world data. Not only do graphing calculators help engage students and deepen their understanding, but they also give students exposure to skills they'll need in advanced mathematics classes and for tests such as the SAT and statewide assessments on which graphing calculators are permitted.

For some investigations, computer software has the power, functionalities, and visual appeal to motivate extended exploration and stimulate new ideas. So, the *Discovering Mathematics* series includes some optional projects, explorations, exercises, and demonstrations using The Geometer's Sketchpad and Fathom Dynamic Statistics software. These programs allow students to visualize, construct, and manipulate geometric figures, algebraic expressions, data displays, and other mathematical objects. In software environments, students encounter intriguing ideas like the Golden Ratio and Moore's Law (algebra), Napoleon's Theorem and a Pythagorean fractal (geometry), and residual plots and construction of conic sections (advanced algebra).

Experience with graphing calculators and educational software will prepare students to appreciate and work with other technologies in their further education and in their careers.

The Difference between *Discovering Mathematics* and Other Approaches

When *Discovering Geometry* was first published in 1989, most mathematics textbooks were strictly "directive," and most still are. In a directive approach, teachers give students a rule or demonstrate a process, usually by reading from the textbook or giving a lecture in front of the class. Students are told facts, shown how to solve a particular kind of problem, and asked to practice working problems very similar to the example. Often, students are expected to memorize rules and processes they don't fully understand. They frequently have trouble solving problems that vary from the example, involve real-world contexts, or ask for justification or explanation. The directive approach works for the students who can understand the demonstrated method, abstract its meaning, and then apply it to new problems. However, students who don't follow the example or who cannot make the mental leap to the reasoning behind the method frequently become bored and frustrated. In spite of exhaustive topical coverage and extended problem sets, textbooks that rely on a directive approach don't overcome the students' lack of involvement in their own learning. These students struggle to connect ideas, advance in understanding, and pass tests. They tend to drop out of mathematics as soon as they can.

Scientific research on brain function and how students learn has shown that students need to make connections with what they already know, and they need to work through ideas to develop meaning and construct their own understanding. These insights have resulted in reform-style or strictly "constructivist" programs that present broad, context-based problems and that require students to invent every step of the solution process. Teachers do not give students mathematical facts but rather help students by asking leading questions and giving them hints. In the strictest of constructivist methods, students write their own mathematical explanations and definitions and keep track of accumulated knowledge themselves in notebooks, their sole reference. Students don't have mathematical examples or definitions in their books. Yet curricula that fully embrace these ideas can be difficult to implement, however sound and involving they may be. Because most teachers need extensive professional development to be able to teach "constructively," such a reform program can be onerous and expensive to implement. The student-centered process invariably leads to an integrated curriculum that doesn't lend itself to discrete courses in algebra and geometry, and it can be difficult to assess just what students know. Parents must also be carefully prepared, and schools need a transition strategy to assimilate students coming from other types of programs.

By contrast to these extremes, the pedagogy of the *Discovering Mathematics* series balances the students' need to be genuinely involved with the instincts and training that teachers have to guide students. The series combines new features developed to address the most important outcomes of educational research, but it retains the familiar features that teachers and students traditionally rely on. The "discovery" approach uses hands-on investigations to engage students and help them to make their own connections to prior learning, to develop and configure their own understanding, and to establish new connections with a growing body of mathematical knowledge. This approach encourages work in cooperative groups in which students can review what they know, attack a problem from different angles, evaluate each other's contributions, articulate their understanding, and learn the value of different approaches. Each group shares its conclusions with the class and hears about other approaches or solutions. This acknowledgment of a student's starting point, and the validation of a student's opinions, creativity, and thinking process, as well as the social setting in which new ideas arise and get considered, all contribute to better attention and motivation at the outset and ultimately to confidence and understanding.

Students who continue through *Discovering Advanced Algebra* attain the comprehensive mathematical knowledge base they need to continue with standard courses in precalculus, calculus, and statistics, prerequisites to many careers in engineering, science, or medicine. They have learned all the same content as their classmates in directively taught courses, but their understanding is deeper, and their problem-solving skills are better. They have had the advantages of group-motivated, self-constructed learning. Moreover, they have had the support of excellent readings and examples that teachers, parents, or mentors can guide them through. (See the section of this guide How Can the Implementation of the *Discovering Mathematics* Series Be Effectively Supported? Also see *Discovering Mathematics: A Guide for Teachers* for more detailed pedagogical explanations and sample lesson presentations.)

Life and Test-Taking Skills

As students investigate interesting mathematics problems in cooperative groups using technology, they improve many skills they'll need outside the classroom.

- Students develop the teamwork skills so valuable in higher education and adult life. Group members learn to brainstorm, plan, define and organize tasks, and communicate their individual and collective results.
- Students improve their skills at test-taking. With the confidence that they understand the concepts and have strategies to develop methods on-the-spot to solve problems on their own, they don't need to memorize lots of possibly meaningless algorithms and are less likely to panic on tests.
- Students improve their reading skills. The student book is designed for students to read. The steps of the investigations are clear and written at a readability level appropriate for most students in the course.
- Students improve their oral and written communication skills, within their cooperative group as well as in the larger group.

- Students improve their active listening skills. They learn to listen to and discuss other students' ideas in groups and in full-class sharing.
- Students learn to exercise critical-thinking skills. They learn to evaluate, compare, and critique various assumptions, methods, and conclusions.
- Students learn organizational skills. In their groups, they organize the tasks they work on together. They also assemble portfolios and keep organized notebooks.
- Students develop independent problem-solving skills. Those who have learned to experiment and to improvise their own conjectures and solutions without being told how to solve a specific problem will become the kind of problem-solvers that employers value.

"Students will be intrigued by [*Discovering Advanced Algebra*'s] exceptional real-world applications. The historical information is very interesting and adds much cultural diversity. Concepts are explored using a variety of methods, including discovery, hands-on, technology, deductive and inductive thinking, and traditional explanation methods, and there is a good balance between skill and application problems in each problem set."

—Jennifer North Morris, Santa Rosa High School, Santa Rosa, CA

Classroom-Based Field Testing

At Key Curriculum Press, in the course of developing the *Discovering Mathematics* series—as well as other core curriculum in precalculus, calculus, statistics, integrated mathematics, and problem solving—we have devoted time, energy, and resources to careful development. We collect information about student and teacher needs, examine educational research, cultivate authors with authentic teaching experience, and produce books and software that foster mathematical understanding and enjoyment.

Key has documented success stories from classroom teachers, administrators, parents, and students from throughout the United States and around the world. These comments and stories emerge through field-test feedback, during in-service training provided by our Professional Development Center, from routine exchanges with our customer service representatives, from the reports of our field sales representatives, and from conversations with mathematics teachers who have paused at Key's booths at conferences. This feedback is forwarded to editors charged with guiding the authors through the revision of materials all the way to the bound book.

The authors of *Discovering Algebra* field-tested their manuscript in their classes during the 1997–1998 school year and made many revisions, sometimes testing the revision that was just made. Their revised manuscript was provided to 31 field-testers in the 1998–1999 school year. These field-testers from around the country (and from two international schools in Malaysia and Hong Kong) came from every type of school (big, small, public, private, rural, suburban, urban) and represented a variety of student constituencies and grade levels. The teachers gave detailed feedback chapter-by-chapter on the reading level of the text and on their classroom management experiences with the investigations, as well as on topical coverage and exercises. During the 1999–2000 school year, 15 schools that had field-tested the material during 1998–1999 participated in a field test of a further-revised version of *Discovering Algebra* and returned feedback on it. The Preliminary Edition of *Discovering Algebra* was made generally available for the 2000–2001 school year. Both the field tests and the preparation of the Preliminary Edition gave authors and editors valuable input for the final hardbound version of *Discovering Algebra* that was published in 2002.

Over 2000 students, including Michael Serra's own students, field-tested the draft versions of *Discovering Geometry* in California, Oregon, Washington, and Pennsylvania before its publication in 1989. For the second edition, 16 teachers and over 1000 students from across the country used the author's revised draft in their classes. Direct input was solicited from students as well as from teachers. Editors had also collected feedback from users who wrote in, unsolicited, or who responded to invitations to comment that appeared in the *Discovering Geometry* Newsletter, from those who attended the user-group meetings at NCTM conferences, and from those who attended our Professional Development Institutes. A great deal of feedback was collected from users on both the first and second editions so that the author had many new ideas to test in his own classroom by the time the third edition was in preparation.

Advanced Algebra Through Data Exploration, the precursor of *Discovering Advanced Algebra,* was tested by 35 teachers and over 2000 students during the 1994–1995 school year. The authors used a draft version in their own classes, gathering important insights for its refinement. As with *Discovering Geometry,* editors had collected feedback from users who wrote in, who attended the user-group meetings at NCTM conferences, or who participated in our Professional Development Institutes. Once again, by the time the new edition was in preparation, the authors and editors had a strong sense of the general direction of response to the book. Nevertheless, we surveyed all users of *Advanced Algebra Through Data Exploration,* and our customer service representatives gathered even more information on the needs of teachers. The authors tested their draft manuscript in their classes during the editorial process that resulted in the current edition of *Discovering Advanced Algebra.*

Review

The books of the *Discovering Mathematics* series, like all our textbooks, have been extensively reviewed. Educational experts across the country and abroad have evaluated its mathematical coverage—its currency, depth, rigor—and its pedagogy. With these professionals, who are usually university educators in mathematics (some involved in teacher training), we ensure that our books are well aligned with current research in teaching methodology and students' cognitive development, new knowledge emerging in the field of mathematics, and content and methods that are newly accessible because of advances in technology.

Teacher reviewers examined the revisions to the manuscript that incorporate field-test feedback. Teacher reviewers gave close attention to the appeal and practicality of investigations and activities for classroom implementation; they also commented on the pacing of lessons and understandability of explanations and examples. The classroom-management advice provided by these reviewers helped the authors and editors shape each student book and was also captured and detailed for each teacher's edition. Reviewers' insights about the kind of support teachers need was the basis for the teaching resources package.

Teachers who are current users of a book under revision are the first people we contact to assist in planning the changes for a new edition. The project editor discusses this teacher input with the authors and other reviewers throughout the development of a new edition.

Equity, or "multicultural," reviewers comment on the social contexts of the mathematics that play out in the applications, art, and student dialogues that populate the books. The equity reviewers' backgrounds, range of experiences, and qualifications give them a sensitivity to ethnic, gender, economic, and lifestyle issues as well as awareness of learning styles based on cultural differences. Their heightened consciousness about the learning environment uniquely positions them to give us important advice about how to reflect society fully and fairly in the books. Respect for disciplines outside mathematics, the depiction of people of all types, ages, and occupations, and the authenticity of multicultural content are their concern. Seeking these reviews early in the process has helped the authors and the editorial and production teams at Key to build mathematical content around data sets that represent daily student life and issues of universal importance, and to appeal to students from all backgrounds and with different talents.

Mathematical accuracy is checked by mathematics professionals (often former teachers) who work through each example and problem and look for missing, partial, or faulty answers. Their work is reviewed by the project editor and is often discussed with the authors. Once the manuscript is turned into book pages, it is a routine part of our process to conduct yet another accuracy check. This check enables us, for example, to ensure that problems that might have been reordered or added late in development have also been checked. Selected portions of the text are subjected to the scrutiny of fact-checkers and science experts as well. This careful attention to detail gives the books of the *Discovering Mathematics* series an extremely low rate of error even in the first printing, reducing frustration for students and teachers.

The names of all reviewers who have consented to be identified, as well as accuracy checkers, are printed in the front matter of our books.

The Editor and Author Interaction

Each book in the series has had a project editor leading its development from the conceptualizing stage through the field test and into the authors' final revision. He or she works with other development editors and provides a sounding board for the author or the author team and guides the changes made in the manuscript from idea to expression in text and art. Each teacher's edition is a major project in itself, headed up by a senior development editor working with a coordinating writer, who is also a mathematics educator with an advanced degree. Yet another editor with a mathematics background manages the development of the rest of the teaching resources package.

All the *Discovering Mathematics* development editors have advanced degrees in mathematics, in applied mathematics such as engineering, or in mathematics education, and all have taught high school mathematics. Their credibility as mathematicians and educators makes them partners with the authors in the process of synthesizing reviews, interpreting field-test feedback and other input, and developing the final product.

Key Curriculum Press staff who worked on *Discovering Mathematics* are named on the copyright pages of the books.

Conclusion

We at Key Curriculum Press have found that engaging students in cooperative groups doing hands-on investigations of interesting, challenging problems that call for using technology results in deeper mathematical understanding and better test performance. No matter how far *Discovering Mathematics* students go in mathematics, they will

- be able to approach new problems with creative strategies.
- know how to interpret data.
- understand logical arguments.
- be able to use mathematical models to help them solve everyday problems.

Finally, more students will actually understand, enjoy, and effectively use mathematics. They will appreciate its power and perhaps some of its beauty. They will be able to capitalize on sound quantitative thinking strategies and abilities to enter and succeed in a variety of fulfilling careers. Perhaps most importantly, they will be much more likely to be informed, productive, and responsible citizens and members of society.

What Is the Scientific Research Base for the *Discovering Mathematics* Series?

Today, educators want programs that are carefully developed to help students learn. They need to know that the programs they are offering are informed by research on how the brain works, how students learn, and what works in the classroom. They want programs that have shown they prepare students to do well on high-stakes tests.

Federal Law PL107-110, commonly known as the No Child Left Behind Act of 2001, has mandated that students use curricula that are based on scientific research and that ensure that students perform well on standardized tests. Because the pedagogy and structure of the *Discovering Mathematics* series is strongly informed by scientific research on how students learn, the series helps them develop deeper understanding. That deeper understanding results in scores for *Discovering Mathematics* students on problem-solving tests that are higher than the scores of students in traditional programs. Moreover, students using *Discovering Mathematics* score as high on skills tests as students in traditional programs do. (See San Diego's Experience in the section Assessing Student Achievement.)

The extensive development of the series by authors who are classroom teachers, informed by professional reviews and by the field-test feedback of many other teachers constitutes an elaborate educational research endeavor. These research components document how the *Discovering Mathematics* series was developed—through careful observation of what works for students and careful implementation of the latest in educational wisdom. (See A Short History of the *Discovering Mathematics* Series earlier in this guide.)

Research on Learning

Research on effective learning environments is consistent with the growing body of research on how brains function. The NCTM (2000) process standards are based on the best of this research. A *Discovering Mathematics* classroom is in harmony with those standards and that research. (See Appendix E for a few references to learning and brain research.)

Research on which the NCTM *Principles and Standards* document is based shows that students learn better if they can relate new content to past experience, actively engage in hands-on experiences that involve trial and error, and see a variety of approaches. Students learn better when they engage in reasoning as they read and articulate ideas, explain why, or ask further questions. Learning is more easily transferred for students who are able to apply mathematics and problem-solving strategies to new situations. These insights are thoroughly integrated into each lesson of the *Discovering Mathematics* series.

CONNECTING NEW LEARNING TO PAST EXPERIENCE

At the cerebral level, learning is a process of creating new neural pathways connecting previously stored concepts and experiences to new ideas. Making sure students have adequate prerequisite experiences before attempting to introduce new information to them is an element of NCTM's Learning Principle and a central characteristic of the *Discovering Mathematics* curriculum. Many investigations give students experience with using and at least intuitively understanding mathematical ideas and concepts before the text introduces terminology or attempts to formalize or generalize the learning. Moreover, the lessons of each book are richly connected to various areas of students' lives, making the learning more engaging and rewarding.

Other kinds of connections are also important to the learning process. Opportunities to share, to discuss, and to write can help students connect their mathematical experiences to their emotional and physical experiences. Such an environment is required by NCTM's Teaching Principle and Communications Standard and facilitated through the *Discovering Mathematics* series. The group work involved in the investigative approach can help students connect with each other on a variety of levels.

ENGAGING IN HANDS-ON EXPERIENCES

The latest research confirms the old statement: "I do and I understand." (See Wolfe 2000, Bransford 1999, and other research references listed on page 49.) In other words, the very complex task of learning depends on using different senses and engaging in trial and error. The more senses students use in a learning experience, the better they will be able to connect and remember the content of that experience. Hands-on group investigations give students opportunities to touch, see, hear, and talk. Students who measure and mark quantities of length and volume are better able to understand the ideas of quantitative measurement and make sense of any relationships they notice or develop in such contexts. Students whose walk is measured by a motion sensor remember the experience and link it to the relationship they discovered between distance and time, as well as to the method they used to work it out. Working on problems they haven't been told how to solve gives learners the chance to try an approach, make adjustments, and try again. Minds develop strong connections through trial and error, as less successful approaches are eliminated. As Jensen (1998) puts it, "In some ways, the worst thing that can happen is for a student to get the right answer immediately." Even if a group is unable to solve a problem, their struggle can make it easier for them to understand once another group or a teacher demonstrates a solution.

SEEING A VARIETY OF APPROACHES

The NCTM Representation Standard recommends using multiple representations of relationships. It was developed in recognition of the fact that, as students come to understand the validity of different approaches to a problem, they have more ways to make connections. For example, an algebraic relationship can be expressed in words, shown in a table, graphed, or written as an equation. A geometric construction can be completed with a compass and straightedge, folded paper, or dynamic geometry software.

NCTM's Problem-Solving Standard encourages the application of a variety of problem-solving strategies. Seeing a variety of approaches helps students develop metacognitive skills: They learn to think about and reflect upon their own thinking as well as that of their classmates. Metacognition is an important part of deepening learning. Brain-function research indicates that metacognition is essential to the assimilation of learning. It's certainly an extremely important, if rarely practiced, life skill.

READING AND ARTICULATING

Communicating and explaining are integral parts of learning for students engaged in mathematics, as emphasized by NCTM's Communication Standard and Reasoning and Proof Standard. Not only is reading an important skill for learning and for life outside the classroom, but also a high level of reading comprehension can strengthen the mathematical skills of reasoning—especially the skill of following a logical argument. When the content of the reading gives students other ways to connect to the mathematical ideas, it is even more valuable. For these reasons, the *Discovering Mathematics* books were developed for students to read, not just for teachers to read and then explain to students.

As students articulate their own ideas and questions orally or in writing, they reinforce their learning. Students express their thoughts during group brainstorming and other discussions, while sharing with the class, and through writing in journals, notebooks, and portfolios. Because *Discovering Mathematics* focuses on understanding, questions are encouraged. Even articulating a question requires a good understanding of some of the ideas involved.

Articulating and listening are skills that go together. Students develop listening skills as they work with other students to solve problems. Good listening skills help students hear and improve their own explanations, thus leading to better understanding.

SOLVING UNIQUE PROBLEMS

Without motivation, deep learning cannot take place. Curiosity, anticipation, and challenge can provide that motivation. All are present when students are given a problem that catches their interest but that they don't know how to solve. For example, a recent survey of *Discovering Algebra* users indicated that one of the students' favorite investigations asks them to simulate a capture-recapture experiment to estimate the number of fish in a lake. The fact that students were not told in advance how to solve this problem helped increase their interest.

Caine and Caine (1997) describe good teaching as ". . . creating hearty (socially and emotionally healthy), exciting, and fascinating experiences that invite learners to work hard while exploring how to do the things that challenge and intrigue them."

Research on Understanding Students' Geometric Thinking: The Van Hiele Levels

Some learning research is specific to particular ideas being learned. While developing *Discovering Geometry,* for example, author Michael Serra found inspiration in the van Hiele model of the development of students' geometric thinking.

According to the van Hieles, students studying geometry progress sequentially through several qualitatively different and increasingly sophisticated levels of geometric thought. Students who are required to study content that is more sophisticated than the developmental level they have attained cannot make sense of that content, and deep understanding loses out to mere memorization.

Here is a simplified summary of the van Hiele levels.

- Level 0 is visual. Students think of shapes as wholes, but they do not attend to geometric properties of shapes. For example, they might distinguish triangles from rectangles without referring to a single property of either shape.
- Level 1 is descriptive and analytic. Students conceptualize shapes as being determined by collections of properties rather than as simply matching visual prototypes. However, they do not see relationships between classes of shapes. For example, a student might contend that a square is not a rectangle.
- Level 2 is abstract and relational. Students can classify shapes hierarchically and give informal arguments to justify their classifications. They still, however, do not grasp that logical deduction is the method for establishing geometric truths.
- Level 3 is formal deduction and proof. Students can formally prove theorems within an axiomatic system. This is the level required for a proof-oriented high school geometry course. Students are not yet ready to think about different axiomatic systems.
- Level 4 is rigor. This higher level of thought is needed to analyze and compare different axiomatic systems. Typically, high school geometry does not reach this level.

Students exploring the Pythagorean Theorem

Over 70% of students begin high school geometry at level 0. Unfortunately, research indicates that only students who enter at level 1 or higher have a good chance of becoming competent with proof—a level 3 activity—by the end of the course. For the many students who are functioning at such low van Hiele levels when they enter high school geometry, traditional courses are dismally ineffective. Indeed, almost 40% of students end the year still below level 1, and only about 30% of the students in courses that teach proof reach a 75% mastery level in writing proofs.

In contrast, the *Discovering Geometry* curriculum is designed to help students move gradually from level 0 to level 1, then to level 2 in the van Hiele hierarchy.

Students are gradually introduced to proof through the process of explaining why. Valid alternative forms of proof, including paragraph proofs and visually easy-to-follow flowchart proofs, appear throughout the book. Only near the end of the course do students see two-column proofs and write formal proofs independently. Thus more students are successful at proofs and those who do not master level 3 (formal deduction and proof) have moved from level 0 or 1 to level 2. The approach to proof taken in *Discovering Geometry* has been reviewed in depth and detail by Michael de Villiers, a world-class leader on reasoning and proof.

Assessing Student Achievement

Verification of learning success is a form of research that follows publication. It provides justification after the fact for the adoption of a particular textbook and supports future adoptions in other schools. Some learner-verification research is necessarily anecdotal, because extensive research is difficult to organize and cumbersome to conduct. Results depend not only on the textbooks being used in the classrooms but also on various other factors: the criteria applied, the testing instruments used, the starting-level of preparedness of students, the learning curve for teachers using a new curriculum, and the willingness of parents, teachers, and administrators to participate in a study.

SAN DIEGO'S EXPERIENCE

Beginning in the fall of 2000, the San Diego school district adopted *Discovering Algebra* for students who had failed a traditional eighth-grade algebra class. They enlisted the help of Texas Instruments Corporation, which provided TI-83 Plus graphing calculators for the classes.

As reported on the Texas Instruments Web site (http://education.ti.com, 2003), the results have been encouraging. The first ninth-grade class to use *Discovering Algebra* experienced a 9% increase in the number of students at or above the 50th percentile on the SAT9, compared to ninth graders in the 1999–2000 school year. These impressive results continued during the 2001–2002 school year, when the number of ninth graders at or above the 50th percentile was 10% higher than the number of ninth graders in 1999–2000.

Teachers saw positive results in their classrooms. Girlie Suero, a teacher who used *Discovering Algebra* at Mira Mesa High School, said, "Investigations and activities such as Walk the Line I and II, Translations of Functions, Sierpiński Triangles, and Equivalent Equations help my students understand that there are at least two ways to approach a problem—algebraically and graphically."

At Kearny High School, Suzanne Barker has witnessed firsthand how *Discovering Algebra* and TI technology have empowered students who might previously have failed mathematics and dropped out of school. "I have seen kids that I had in Algebra explorations two years ago explain how to do transformations to other students in my Intermediate class," she says. "In fact, some of these students could explain and equate rate of change and slope to honors precalculus students. When you see something like that, you know the hard work to help students synthesize and discuss concepts is worth it."

RESEARCH ON *DISCOVERING ALGEBRA*'S PRELIMINARY EDITION

To learn how a wide range of students using *Discovering Algebra* would perform on standardized tests, Key engaged Heller Research Associates of Oakland, California, to evaluate the algebra achievement of students using the Preliminary Edition of *Discovering Algebra*. Their experimental-control study showed several significant results.

The students using *Discovering Algebra* scored as high as or higher than other students on three tests. *Discovering Algebra* students scored significantly higher on the two test forms that focused on conceptual aspects of algebra (one based on a *Discovering Algebra* test and one on the Massachusetts Comprehensive Assessment System Mathematics exam) and reached the same level as their non–*Discovering Algebra* counterparts on test forms that emphasized basic skills and procedures (based on the Texas Assessment of Academic Skills). *Discovering Algebra* was found to benefit students of all ability levels. The study also gave reason to conclude that there may be additional benefits for the strongest students.

Results also suggest that, even though *Discovering Algebra* addresses a set of skills and abilities different from more traditional texts, it supports students in learning more basic skills and procedures and clearly does not detract from basic skill development. The study also established that the higher scores of students who studied with *Discovering Algebra* were not attributable to the use of graphing calculators, although the classroom questionnaires indicate that graphing calculators were more available and used more often in *Discovering Algebra* classes than in classrooms using other texts.

Throughout the analyses of test data, whether of the total sample or of the subset of students in regular classes, *Discovering Algebra* students scored at least as well as non–*Discovering Algebra* students on all test forms, and significantly higher on the two tests with a more conceptual emphasis.

Key Curriculum Press is conducting further controlled studies of the effectiveness of the *Discovering Mathematics* curriculum. We are confident that these studies will also show that *Discovering Mathematics* students do well on tests. This confidence comes from knowing that *Discovering Mathematics* is based on research into how people learn and comes from classroom teachers who see the results in the engagement and achievement of their own students.

The skill and experience of the editorial staff at Key Curriculum Press, along with Key's authors and field-test teachers and the many reviewers and consultants combine to make *Discovering Mathematics* an optimal curriculum.

How Can the Implementation of the *Discovering Mathematics* Series Be Effectively Supported?

So far, most of this guide has described the benefits of *Discovering Mathematics* for students. It has shown that they learn the mathematics needed for understanding, for good test results, and for better career options and life skills, thanks to the curriculum's research basis and its careful development.

As an administrator, you have other concerns as well: about supporting teachers, about satisfying parents and other members of the community, and about the general logistics of scheduling, providing technology, and articulation with feeder schools.

Supporting Teachers

The *Discovering Mathematics* books accommodate the traditional course offerings and have familiar grouping of examples and exercises. However, their emphasis on student investigations in cooperative groups using technology rather than on teacher presentations may make some teachers apprehensive. Some might feel that their ways of teaching are being criticized and that they will be required to make significant and uncomfortable changes to the way they teach. Actually, as the teachers will discover, a range of classroom practices and teaching styles can be accommodated in the *Discovering Mathematics* curriculum without compromising its main advantages. (For a sample lesson from each book of the series, see *Discovering Mathematics: A Guide for Teachers.*)

Teaching an investigation-based approach will alter the way teachers think about their roles in the learning process. Although teachers who have used lecture as their chief pedagogic method will need to make greater adjustments than teachers who have always promoted discussion and collaboration, all teachers have experience with rhetorical or Socratic questioning. That is, teachers have always expected students to ponder and figure out some things before they provide students with the answers, they have expected students to show their work, and they have witnessed the power of peer coaching.

Thanks to extensive teacher support, teachers will adapt readily and find the discovery approach a natural way to teach. Most teachers will welcome the changes they learn about and can make without extensive retraining. Many will recognize that they have finally found books and resources that will help them teach the way they have always wanted to, without the need to create their own materials or do research on their own. Essays found at the beginning of each teacher's edition and in the book *Discovering Mathematics: A Guide for Teachers* help teachers as they implement an investigation-based classroom. These books and resources at **www.keypress.com/DM** —including the chance to exchange ideas with other teachers—give them support as they develop professionally.

The wrap-around commentary in the teacher's editions embeds professional development in teachers' daily planning and classroom practice. For example, it reminds teachers of the importance of giving students a chance to make mistakes and encourages teachers to ask questions for which they do not have the answer such as "Why did you decide to take this approach?" It gives extensive advice regarding the use of cooperative groups. As it talks through the investigation steps and the sharing of ideas, the commentary guides the teacher to facilitate students' pooling of information and drawing jointly agreed-upon conclusions. Lesson objectives and a section called Closing the Lesson help teachers make sure that the important mathematics is not lost in the excitement of activities.

Teachers who need technology help will find informative, clearly written support materials. *Discovering Algebra* and *Discovering Advanced Algebra* have calculator notes keyed to the lessons for the most popular Texas Instruments and Casio graphing calculators. Teachers who have access to computers with Sketchpad and Fathom software can consult *Discovering Geometry Demonstrations with The Geometer's Sketchpad, Discovering Advanced Algebra Demonstrations with Fathom and The Geometer's Sketchpad,* and *Discovering Geometry with The Geometer's Sketchpad.* Each book and CD includes software instructions and all the sample files needed to complete the Fathom and Sketchpad explorations in the student books, to replace a *Discovering Geometry* lesson with the same lesson using Sketchpad, or to demonstrate a relationship in geometry, algebra, or statistics. These ancillaries also include ideas and instructions for using the dynamic software programs as demonstrations in other lessons or as replacement lessons for *Discovering Geometry.*

For teachers who want to improve their teaching skills further, training and support are available through workshops offered by Key Curriculum Press's Professional Development Center.

Your support is essential as teachers work with any new curriculum. You'll also want to consider changing the way teachers are evaluated. Traditional evaluation methods focus on teacher behavior, but in the *Discovering Mathematics* classroom it would be much more appropriate to focus on student behavior and, eventually, on student performance.

Building Parent Support

As with any other mathematics curriculum, parents can play a large part in the success of the program and in their student's success at using it. Parents who come to understand the strengths of this curriculum will be enthusiastic about the discovery-learning environment. Parents will welcome what students are gaining when they see students look for patterns, make conjectures, try different models, and talk about what they are doing and learning. Once parents understand the value of building understanding, they will see that *Discovering Mathematics* is teaching essential life skills.

However, because the *Discovering Mathematics* curriculum is different from most traditional programs or textbooks that parents used in school, concerned parents may have questions about the investigative approach and its effectiveness. Experience shows that a number of steps can be taken to help parents overcome their discomfort with a curriculum differing from the one they experienced as students.

In some schools, special "back to school" nights can attract many parents. These can be especially effective if parents actually work through a hands-on lesson as students would: in small groups, explaining their ideas to each other, and seeing the value of multiple approaches. From their own learning, parents can appreciate how the program was designed to help all students develop a deep understanding of mathematical concepts. After such an activity, they'll be open to seeing overviews of the books and hearing that all the mathematical ideas in a traditional course appear in *Discovering Mathematics,* so that students will be fully prepared for precalculus and advanced placement calculus and statistics courses. Parents will also realize the advantages students have when they learn to work in teams to approach problems they haven't previously been told how to solve. More ideas about planning a night to introduce parents to *Discovering Mathematics* can be found at www.keypress.com/backtoschool.

A fall newsletter explaining the program can also be useful. It might even include advice about what kinds of calculators parents might consider purchasing for their students at home. Some schools have encouraged parents to watch a TIMSS 1999 Video Study available from the National Center for Education Statistics. It will help if you have at least several excellent articles and a couple of good Web sites to which to refer concerned parents. Some schools implementing new mathematics curricula have actually provided an evening tutor just for parents.

Parents are sometimes concerned that they can't help their students with their homework in this curriculum. Refer these parents to the Web site created for this purpose at www.keymath.com. There, they can find explanations of the mathematics of each chapter and a summary problem that they can talk about with their student as they work through the chapter. The same parent Web site provides information on resources available for extra skills practice or to help their student catch up when absent, including *Condensed Lessons for Make-Up Work* (also available in Spanish).

Parents may be concerned that when students work cooperatively, one or two students in a group will do all the work. One of the teacher's responsibilities is to make sure that each and every student plays an active role in the group and is held accountable for his or her own work and participation as well as the group's result. It might help to let parents know that the composition of groups will be changed regularly throughout the year, and that teachers do not plan to give the same term grade to an entire group. More resources for communicating with parents about such concerns can be found in *Discovering Mathematics: A Guide for Teachers* and at www.keypress.com/DM/teachersguide.

You will also want to be able to justify the curriculum to others in the community. Business leaders will be happy to know that many of the skills taught in *Discovering Mathematics* are those they value in their employees: skills of calculating, measurement, elementary statistics, reading, writing, problem solving, and teamwork. Accrediting associations will also be pleased that you have a problem-based mathematics curriculum consonant with scientific research on brain function and learning.

Logistics

A major concern of administrators is the availability of technology needed to support the *Discovering Mathematics* curriculum. Students using *Discovering Algebra* are not required to own their own calculators, but classroom sets are essential. Students using *Discovering Advanced Algebra* need access to graphing calculators for homework as well. Various grants are available to help schools purchase calculators and software site licenses.

Another of your concerns might be how *Discovering Mathematics* fits into a special schedule, such as blocks or trimesters. Pacing guides for various schedules appear at the beginning of each teacher's edition.

APPENDIX A The *Discovering Mathematics* Series Authors

Author Michael Serra was teaching geometry at Washington High School in San Francisco, California, when he began developing geometry curricula. For ten years he created and used his own materials, applying the information he had gathered from conferences about how students learn geometry. Learning theory indicated that students should start with the concrete and move to the abstract, but geometry textbooks started with abstract postulates and required students to re-create proofs they often did not understand.

Michael Serra

During that same time, Serra was working on a Research in Industry Grant where he heard repeatedly that skills valued by industry included the ability to communicate orally and in writing and the ability to work in groups. He began to develop these skills in his students. His teaching experience became the basis for *Discovering Geometry: An Inductive Approach.* Michael Serra is a noted speaker on inductive geometry and cooperative learning at local, state, and national meetings.

Authors Jerald Murdock and Ellen and Eric Kamischke began working together at Interlochen Arts Academy in Interlochen, Michigan. They used graphing calculators to teach algebra when that technology first became available for classrooms. Partially supported by grant money from the National Science Foundation, they began doing investigations with their own students using real data and hands-on experiments. The result was published by Key Curriculum Press as *Advanced Algebra Through Data Exploration: A Graphing Calculator Approach.* It has been revised to become *Discovering Advanced Algebra: An Investigative Approach.* This same group of teachers authored *Discovering Algebra: An Investigative Approach.*

Jerald Murdock working with students.

Jerald Murdock is a Presidential Awardee for Excellence in Mathematics Teaching and a Woodrow Wilson Fellow. He has taught in both public and private high schools and is an experienced speaker and workshop leader.

Ellen Kamischke working with two students.

Ellen Kamischke holds degrees in mathematics and physics. She enjoys finding ways to incorporate the arts in her teaching and making the learning of mathematics both challenging and exciting for her students. She is a workshop leader and a frequent presenter at regional and national mathematics conferences on topics ranging from writing in mathematics to algebra and calculus.

Eric Kamischke

Eric Kamischke is also a Woodrow Wilson Fellow. A former chemistry teacher, he's an expert in classroom technology and uses it for lab investigations in his mathematics teaching and in his presentations to teachers. Eric is an experienced presenter who has shared his technological expertise nationwide.

APPENDIX B Tables of Contents

DISCOVERING ALGEBRA

DISCOVERING GEOMETRY

DISCOVERING ADVANCED ALGEBRA

APPENDIX C Features of the *Discovering Mathematics* Series

Teacher feedback and suggestions helped shape the instructional design of *Discovering Algebra,* creating the look and feel that would continue in *Discovering Geometry* and *Discovering Advanced Algebra.* Text, graphics, and art facilitate navigation through the lessons, leading students from the introductions to the investigations, connections, review, and more. The wrap-around style of the teacher's edition pages gives teachers readable narratives and comprehensive support with both the mathematical content and the pedagogy.

The following pages call out some of the helpful features of the student and teacher books. For a more complete overview of the features of the books in the *Discovering Mathematics* series, call 800.995.MATH (800.955.6284) and request a copy of the *Discovering Mathematics brochure.*

INVESTIGATIONS promote responsibility for learning and lead to important insights. Some lessons have two or three investigations, so teachers can opt to have students work through each investigation or have groups work on different investigations, pooling their results in class discussion. **NUMBERED STEPS** help students navigate the investigation independently and help teachers gauge group progress through the work.

LESSONS are organized around clear objectives. Opening text orients students in the content sequence and often relates the topic to real-world experience.

PLANNING helps teachers complete lesson plans, gather materials, and prepare worksheets, transparencies, or calculator notes. **LESSON OUTLINE** helps teachers structure the class period and lets them know whether they should plan to spend two days on the lesson.

TEACHING gives practical help in guiding the investigation, facilitating student sharing, evaluating progress, and explaining new mathematical ideas.

ONE STEP is for classes experienced at investigating. It is an alternative to the more closely guided investigation steps in the book.

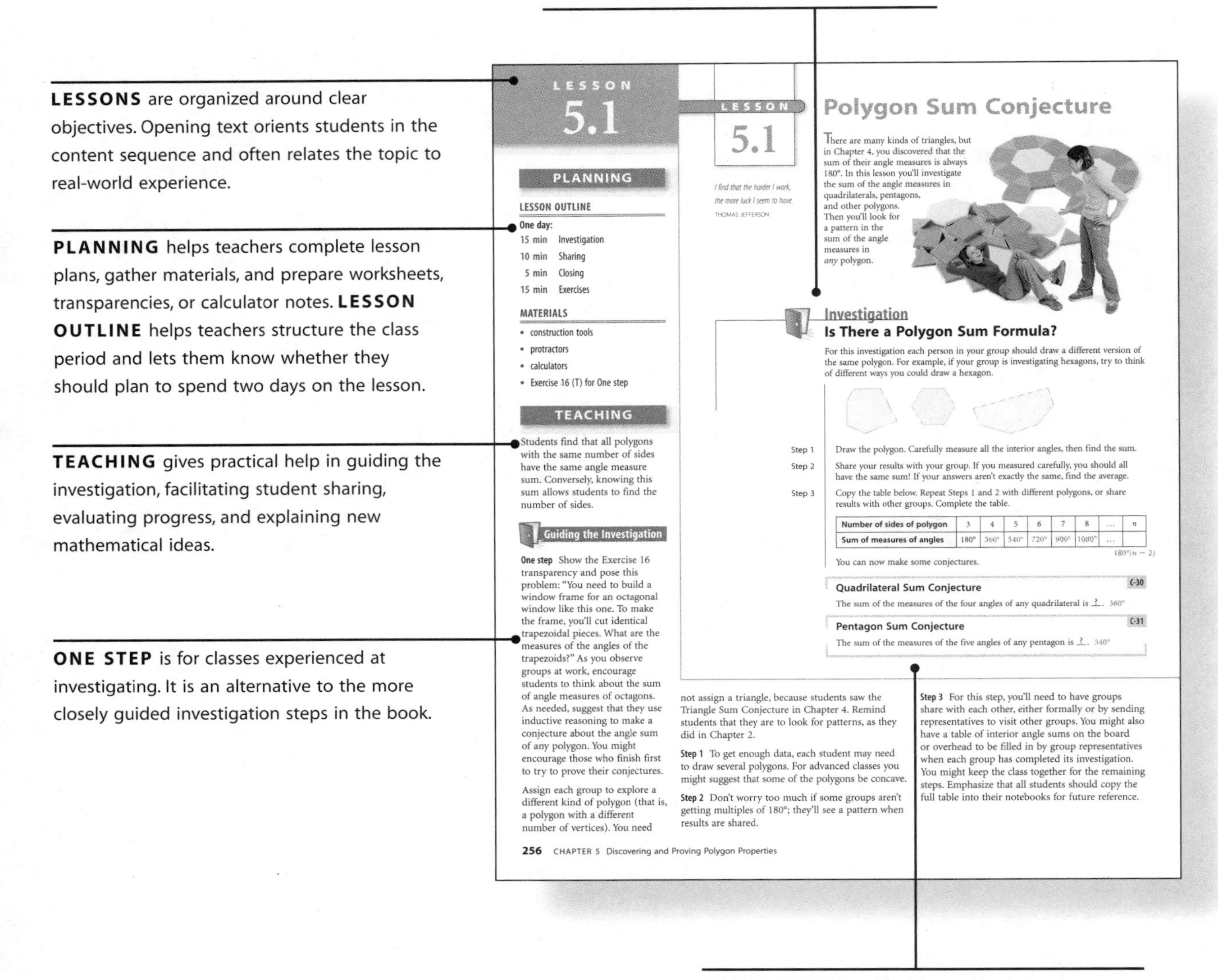

LESSON 5.1

PLANNING

LESSON OUTLINE

One day:
15 min Investigation
10 min Sharing
5 min Closing
15 min Exercises

MATERIALS

- construction tools
- protractors
- calculators
- Exercise 16 (T) for One step

TEACHING

Students find that all polygons with the same number of sides have the same angle measure sum. Conversely, knowing this sum allows students to find the number of sides.

Guiding the Investigation

One step Show the Exercise 16 transparency and pose this problem: "You need to build a window frame for an octagonal window like this one. To make the frame, you'll cut identical trapezoidal pieces. What are the measures of the angles of the trapezoids?" As you observe groups at work, encourage students to think about the sum of angle measures of octagons. As needed, suggest that they use inductive reasoning to make a conjecture about the angle sum of any polygon. You might encourage those who finish first to try to prove their conjectures.

Assign each group to explore a different kind of polygon (that is, a polygon with a different number of vertices). You need not assign a triangle, because students saw the Triangle Sum Conjecture in Chapter 4. Remind students that they are to look for patterns, as they did in Chapter 2.

Step 1 To get enough data, each student may need to draw several polygons. For advanced classes you might suggest that some of the polygons be concave.

Step 2 Don't worry too much if some groups aren't getting multiples of 180°; they'll see a pattern when results are shared.

Step 3 For this step, you'll need to have groups share with each other, either formally or by sending representatives to visit other groups. You might also have a table of interior angle sums on the board or overhead to be filled in by group representatives when each group has completed its investigation. You might keep the class together for the remaining steps. Emphasize that all students should copy the full table into their notebooks for future reference.

LESSON 5.1

I find that the harder I work, the more luck I seem to have.
THOMAS JEFFERSON

Polygon Sum Conjecture

There are many kinds of triangles, but in Chapter 4, you discovered that the sum of their angle measures is always 180°. In this lesson you'll investigate the sum of the angle measures in quadrilaterals, pentagons, and other polygons. Then you'll look for a pattern in the sum of the angle measures in *any* polygon.

Investigation

Is There a Polygon Sum Formula?

For this investigation each person in your group should draw a different version of the same polygon. For example, if your group is investigating hexagons, try to think of different ways you could draw a hexagon.

Step 1 Draw the polygon. Carefully measure all the interior angles, then find the sum.

Step 2 Share your results with your group. If you measured carefully, you should all have the same sum! If your answers aren't exactly the same, find the average.

Step 3 Copy the table below. Repeat Steps 1 and 2 with different polygons, or share results with other groups. Complete the table.

Number of sides of polygon	3	4	5	6	7	8	...	n
Sum of measures of angles	180°	360°	540°	720°	900°	1080°	...	$180°(n-2)$

You can now make some conjectures.

Quadrilateral Sum Conjecture C-30

The sum of the measures of the four angles of any quadrilateral is ?. 360°

Pentagon Sum Conjecture C-31

The sum of the measures of the five angles of any pentagon is ?. 540°

256 CHAPTER 5 Discovering and Proving Polygon Properties

CONJECTURES guide *Discovering Geometry* students' conclusions into clear mathematical language and stock the student's evolving knowledge framework with solid content.

SHARING IDEAS guides teachers in working with students as they present their results from the investigation and in helping students understand the mathematics.

PAUSE RULES mark good places to stop for discussions or to break for the next class period.

FINISH RULES signal the end of the investigation.

Help is keyed to the **STEPS** of the investigation.

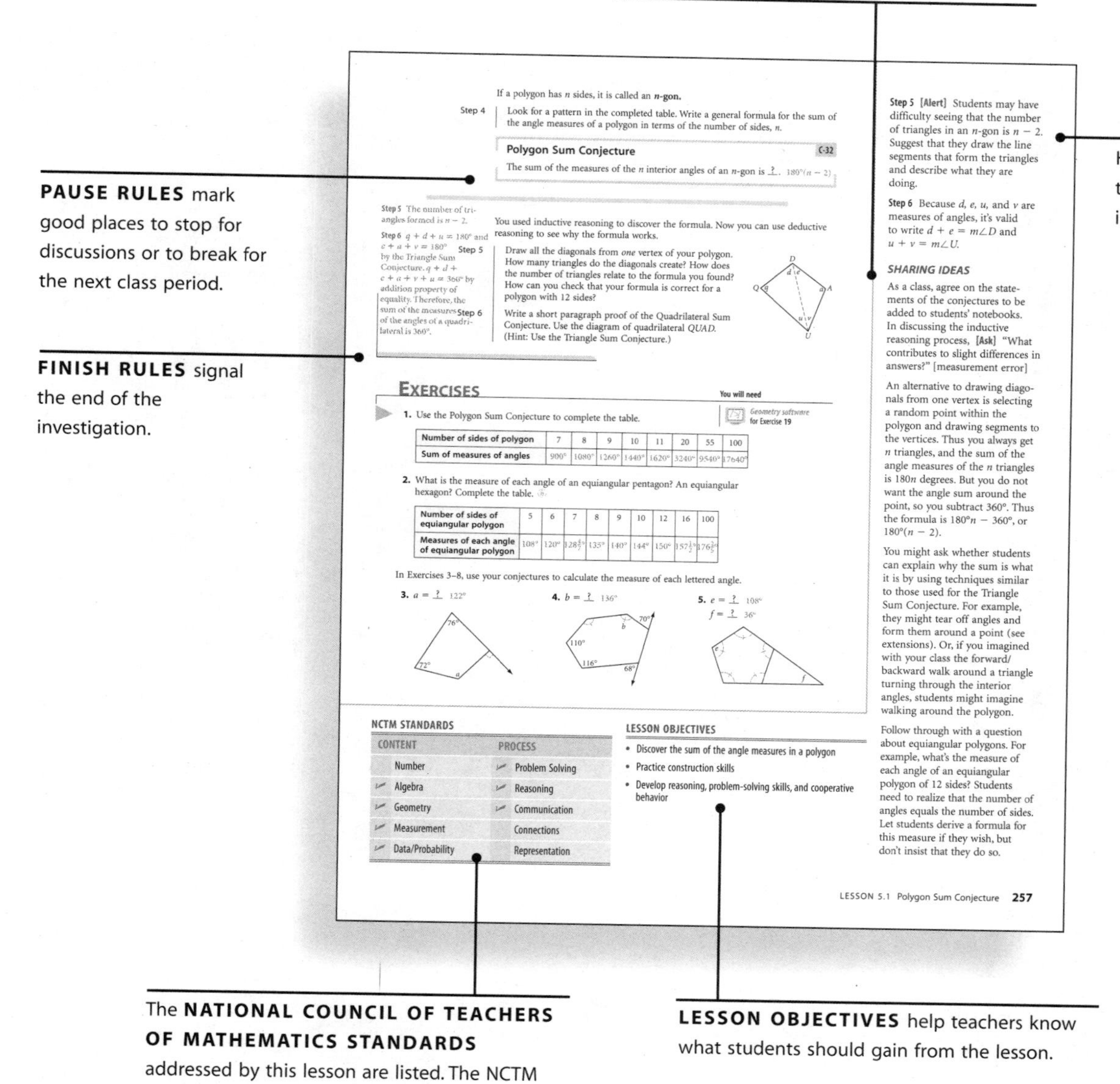

If a polygon has n sides, it is called an ***n*-gon.**

Step 4 Look for a pattern in the completed table. Write a general formula for the sum of the angle measures of a polygon in terms of the number of sides, n.

Polygon Sum Conjecture C-32

The sum of the measures of the n interior angles of an n-gon is ?. $180°(n - 2)$

Step 5 The number of triangles formed is $n - 2$.

Step 6 $q + d + u = 180°$ and $e + a + v = 180°$ by the Triangle Sum Conjecture. $q + d + e + a + v + u = 360°$ by addition property of equality. Therefore, the sum of the measures of the angles of a quadrilateral is 360°.

You used inductive reasoning to discover the formula. Now you can use deductive reasoning to see why the formula works.

Step 5 Draw all the diagonals from *one* vertex of your polygon. How many triangles do the diagonals create? How does the number of triangles relate to the formula you found? How can you check that your formula is correct for a polygon with 12 sides?

Step 6 Write a short paragraph proof of the Quadrilateral Sum Conjecture. Use the diagram of quadrilateral *QUAD*. (Hint: Use the Triangle Sum Conjecture.)

EXERCISES

You will need

Geometry software for Exercise 19

1. Use the Polygon Sum Conjecture to complete the table.

Number of sides of polygon	7	8	9	10	11	20	55	100
Sum of measures of angles	900°	1080°	1260°	1440°	1620°	3240°	9540°	17640°

2. What is the measure of each angle of an equiangular pentagon? An equiangular hexagon? Complete the table.

Number of sides of equiangular polygon	5	6	7	8	9	10	12	16	100
Measures of each angle of equiangular polygon	108°	120°	$128\frac{4}{7}°$	135°	140°	144°	150°	$157\frac{1}{2}°$	$176\frac{2}{5}°$

In Exercises 3–8, use your conjectures to calculate the measure of each lettered angle.

3. $a =$? 122°

4. $b =$? 136°

5. $e =$? 108°
$f =$? 36°

Step 5 [Alert] Students may have difficulty seeing that the number of triangles in an n-gon is $n - 2$. Suggest that they draw the line segments that form the triangles and describe what they are doing.

Step 6 Because d, e, u, and v are measures of angles, it's valid to write $d + e = m\angle D$ and $u + v = m\angle U$.

SHARING IDEAS

As a class, agree on the statements of the conjectures to be added to students' notebooks. In discussing the inductive reasoning process, **[Ask]** "What contributes to slight differences in answers?" [measurement error]

An alternative to drawing diagonals from one vertex is selecting a random point within the polygon and drawing segments to the vertices. Thus you always get n triangles, and the sum of the angle measures of the n triangles is $180n$ degrees. But you do not want the angle sum around the point, so you subtract 360°. Thus the formula is $180°n - 360°$, or $180°(n - 2)$.

You might ask whether students can explain why the sum is what it is by using techniques similar to those used for the Triangle Sum Conjecture. For example, they might tear off angles and form them around a point (see extensions). Or, if you imagined with your class the forward/backward walk around a triangle turning through the interior angles, students might imagine walking around the polygon.

Follow through with a question about equiangular polygons. For example, what's the measure of each angle of an equiangular polygon of 12 sides? Students need to realize that the number of angles equals the number of sides. Let students derive a formula for this measure if they wish, but don't insist that they do so.

NCTM STANDARDS

CONTENT		PROCESS	
	Number	✓	Problem Solving
✓	Algebra	✓	Reasoning
✓	Geometry	✓	Communication
✓	Measurement		Connections
✓	Data/Probability		Representation

LESSON OBJECTIVES

- Discover the sum of the angle measures in a polygon
- Practice construction skills
- Develop reasoning, problem-solving skills, and cooperative behavior

LESSON 5.1 Polygon Sum Conjecture **257**

The **NATIONAL COUNCIL OF TEACHERS OF MATHEMATICS STANDARDS** addressed by this lesson are listed. The NCTM *Principles* as they apply to the entire book are discussed in the front matter.

LESSON OBJECTIVES help teachers know what students should gain from the lesson.

EXAMPLES, often set in real-world scenarios, show how to identify important information, how to define variables and write equations, how to make graphs and tables, and how to present solutions clearly.

TEACHING notes suggest alternative approaches.

SOLUTIONS are provided for each example. They are fully worked out computationally with reasons that justify the process. Notation conventions are clearly explained. Results are interpreted back into the real-world context.

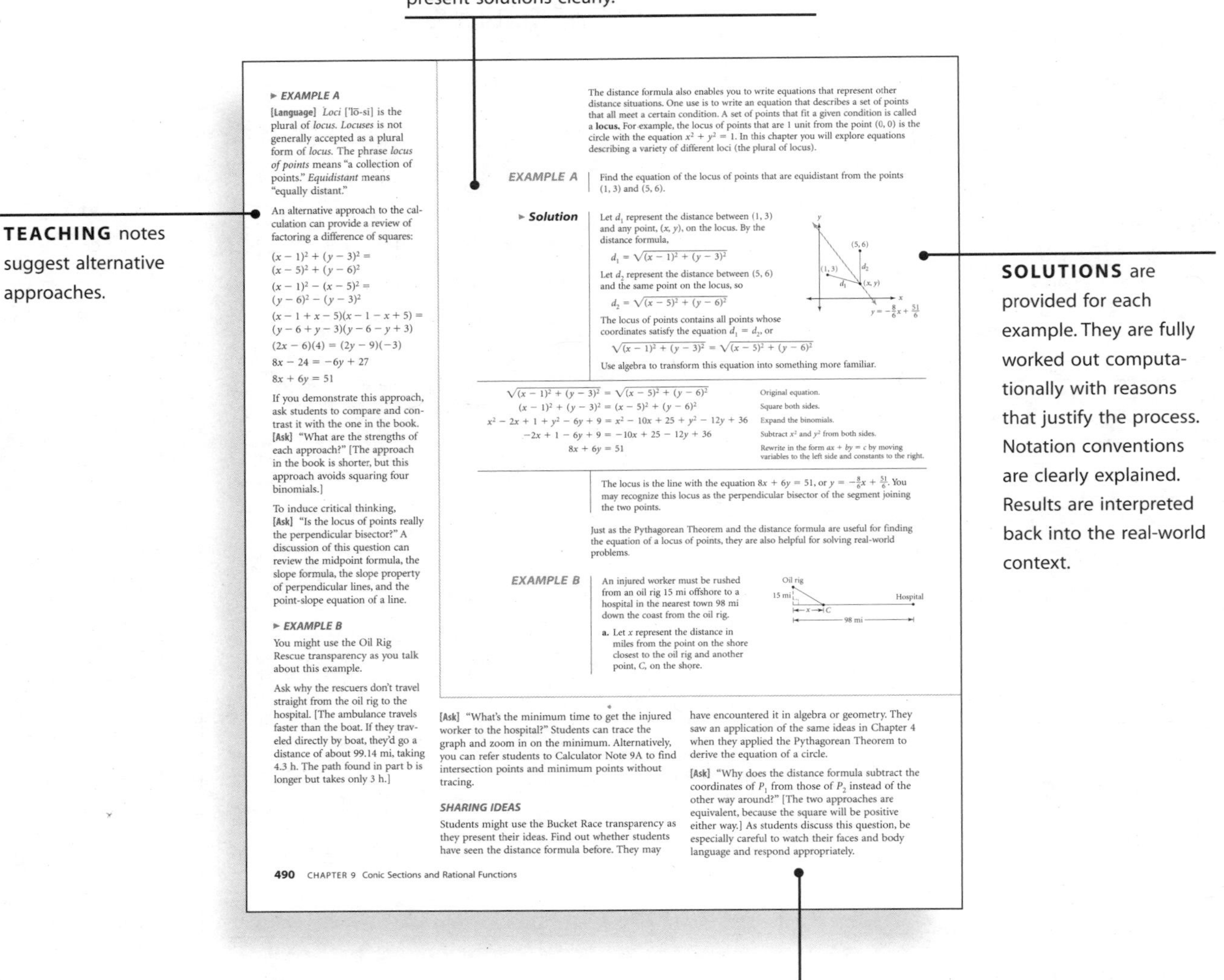

▸ *EXAMPLE A*

[Language] *Loci* ['lō-si] is the plural of *locus*. *Locuses* is not generally accepted as a plural form of *locus*. The phrase *locus of points* means "a collection of points." *Equidistant* means "equally distant."

An alternative approach to the calculation can provide a review of factoring a difference of squares:

$(x-1)^2 + (y-3)^2 =$
$(x-5)^2 + (y-6)^2$

$(x-1)^2 - (x-5)^2 =$
$(y-6)^2 - (y-3)^2$

$(x-1+x-5)(x-1-x+5) =$
$(y-6+y-3)(y-6-y+3)$

$(2x-6)(4) = (2y-9)(-3)$

$8x - 24 = -6y + 27$

$8x + 6y = 51$

If you demonstrate this approach, ask students to compare and contrast it with the one in the book. **[Ask]** "What are the strengths of each approach?" [The approach in the book is shorter, but this approach avoids squaring four binomials.]

To induce critical thinking, **[Ask]** "Is the locus of points really the perpendicular bisector?" A discussion of this question can review the midpoint formula, the slope formula, the slope property of perpendicular lines, and the point-slope equation of a line.

▸ *EXAMPLE B*

You might use the Oil Rig Rescue transparency as you talk about this example.

Ask why the rescuers don't travel straight from the oil rig to the hospital. [The ambulance travels faster than the boat. If they traveled directly by boat, they'd go a distance of about 99.14 mi, taking 4.3 h. The path found in part b is longer but takes only 3 h.]

The distance formula also enables you to write equations that represent other distance situations. One use is to write an equation that describes a set of points that all meet a certain condition. A set of points that fit a given condition is called a **locus.** For example, the locus of points that are 1 unit from the point (0, 0) is the circle with the equation $x^2 + y^2 = 1$. In this chapter you will explore equations describing a variety of different loci (the plural of locus).

EXAMPLE A Find the equation of the locus of points that are equidistant from the points (1, 3) and (5, 6).

▸ ***Solution*** Let d_1 represent the distance between (1, 3) and any point, (x, y), on the locus. By the distance formula,

$$d_1 = \sqrt{(x-1)^2 + (y-3)^2}$$

Let d_2 represent the distance between (5, 6) and the same point on the locus, so

$$d_2 = \sqrt{(x-5)^2 + (y-6)^2}$$

The locus of points contains all points whose coordinates satisfy the equation $d_1 = d_2$, or

$$\sqrt{(x-1)^2 + (y-3)^2} = \sqrt{(x-5)^2 + (y-6)^2}$$

Use algebra to transform this equation into something more familiar.

$\sqrt{(x-1)^2 + (y-3)^2} = \sqrt{(x-5)^2 + (y-6)^2}$	Original equation.
$(x-1)^2 + (y-3)^2 = (x-5)^2 + (y-6)^2$	Square both sides.
$x^2 - 2x + 1 + y^2 - 6y + 9 = x^2 - 10x + 25 + y^2 - 12y + 36$	Expand the binomials.
$-2x + 1 - 6y + 9 = -10x + 25 - 12y + 36$	Subtract x^2 and y^2 from both sides.
$8x + 6y = 51$	Rewrite in the form $ax + by = c$ by moving variables to the left side and constants to the right.

The locus is the line with the equation $8x + 6y = 51$, or $y = -\frac{8}{6}x + \frac{51}{6}$. You may recognize this locus as the perpendicular bisector of the segment joining the two points.

Just as the Pythagorean Theorem and the distance formula are useful for finding the equation of a locus of points, they are also helpful for solving real-world problems.

EXAMPLE B An injured worker must be rushed from an oil rig 15 mi offshore to a hospital in the nearest town 98 mi down the coast from the oil rig.

a. Let x represent the distance in miles from the point on the shore closest to the oil rig and another point, C, on the shore.

[Ask] "What's the minimum time to get the injured worker to the hospital?" Students can trace the graph and zoom in on the minimum. Alternatively, you can refer students to Calculator Note 9A to find intersection points and minimum points without tracing.

SHARING IDEAS

Students might use the Bucket Race transparency as they present their ideas. Find out whether students have seen the distance formula before. They may have encountered it in algebra or geometry. They saw an application of the same ideas in Chapter 4 when they applied the Pythagorean Theorem to derive the equation of a circle.

[Ask] "Why does the distance formula subtract the coordinates of P_1 from those of P_2 instead of the other way around?" [The two approaches are equivalent, because the square will be positive either way.] As students discuss this question, be especially careful to watch their faces and body language and respond appropriately.

490 CHAPTER 9 Conic Sections and Rational Functions

TEACHING notes suggest good questions to ask as the class reviews the examples or shares ideas.

GRAPHING CALCULATOR use in *Discovering Algebra* and *Discovering Advanced Algebra* is modeled in Examples, Investigations, and Exercises whenever advisable and practical. Screen captures demonstrate the functions available to students through technology. Separate Calculator Notes help keep the text page uncluttered, so that mathematics remains paramount over keystroke detail.

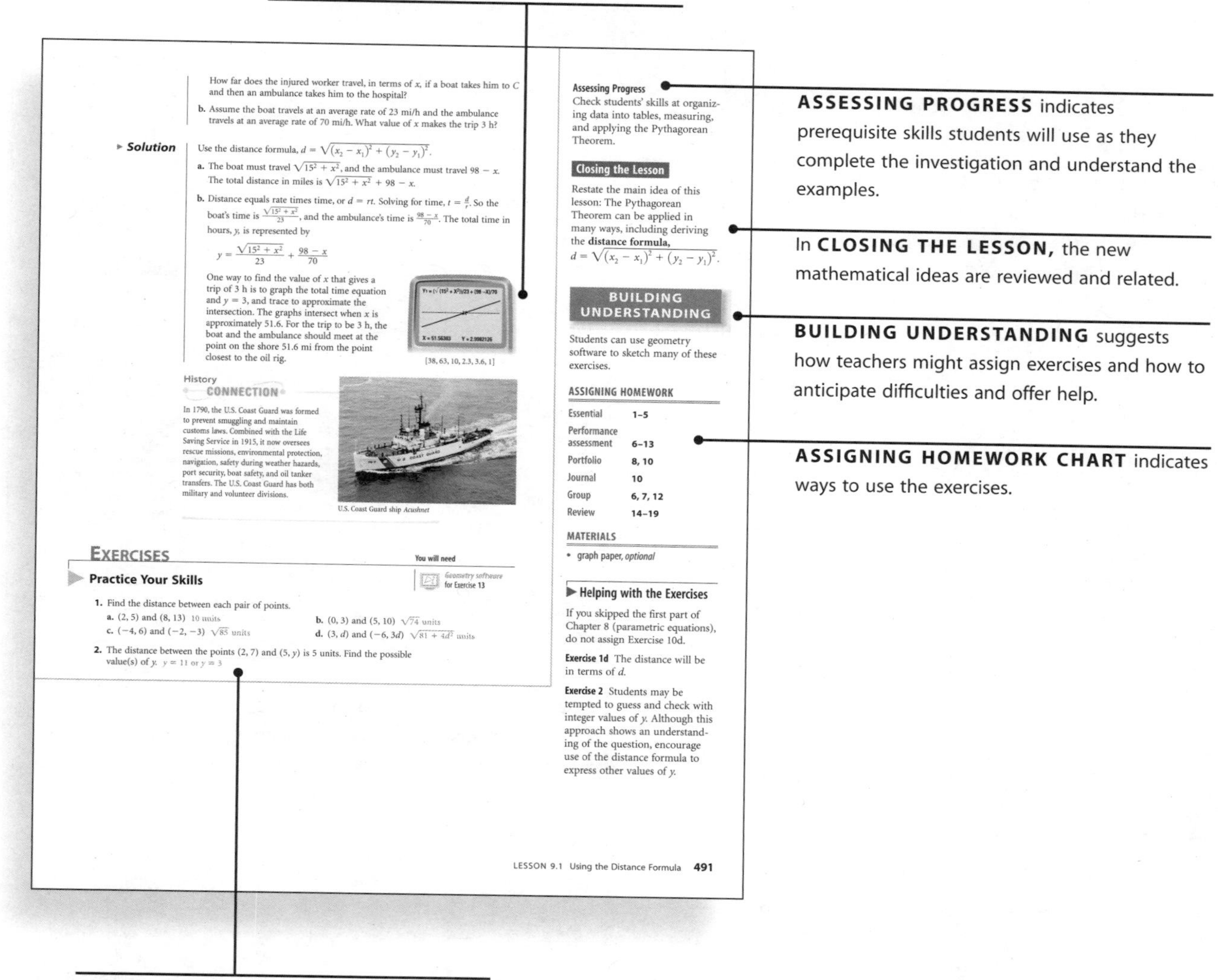

How far does the injured worker travel, in terms of x, if a boat takes him to C and then an ambulance takes him to the hospital?

b. Assume the boat travels at an average rate of 23 mi/h and the ambulance travels at an average rate of 70 mi/h. What value of x makes the trip 3 h?

▸ Solution

Use the distance formula, $d = \sqrt{(x_2 - x_1)^2 + (y_2 - y_1)^2}$.

a. The boat must travel $\sqrt{15^2 + x^2}$, and the ambulance must travel $98 - x$. The total distance in miles is $\sqrt{15^2 + x^2} + 98 - x$.

b. Distance equals rate times time, or $d = rt$. Solving for time, $t = \frac{d}{r}$. So the boat's time is $\frac{\sqrt{15^2 + x^2}}{23}$, and the ambulance's time is $\frac{98 - x}{70}$. The total time in hours, y, is represented by

$$y = \frac{\sqrt{15^2 + x^2}}{23} + \frac{98 - x}{70}$$

One way to find the value of x that gives a trip of 3 h is to graph the total time equation and $y = 3$, and trace to approximate the intersection. The graphs intersect when x is approximately 51.6. For the trip to be 3 h, the boat and the ambulance should meet at the point on the shore 51.6 mi from the point closest to the oil rig.

[38, 63, 10, 2.3, 3.6, 1]

History
CONNECTION

In 1790, the U.S. Coast Guard was formed to prevent smuggling and maintain customs laws. Combined with the Life Saving Service in 1915, it now oversees rescue missions, environmental protection, navigation, safety during weather hazards, port security, boat safety, and oil tanker transfers. The U.S. Coast Guard has both military and volunteer divisions.

U.S. Coast Guard ship *Acushnet*

EXERCISES

You will need

Geometry software for Exercise 13

Practice Your Skills

1. Find the distance between each pair of points.
   - **a.** (2, 5) and (8, 13) 10 units
   - **b.** (0, 3) and (5, 10) $\sqrt{74}$ units
   - **c.** (−4, 6) and (−2, −3) $\sqrt{85}$ units
   - **d.** (3, d) and (−6, 3d) $\sqrt{81 + 4d^2}$ units
2. The distance between the points (2, 7) and (5, y) is 5 units. Find the possible value(s) of y. $y = 11$ or $y = 3$

Assessing Progress
Check students' skills at organizing data into tables, measuring, and applying the Pythagorean Theorem.

Closing the Lesson

Restate the main idea of this lesson: The Pythagorean Theorem can be applied in many ways, including deriving the **distance formula,** $d = \sqrt{(x_2 - x_1)^2 + (y_2 - y_1)^2}$.

BUILDING UNDERSTANDING

Students can use geometry software to sketch many of these exercises.

ASSIGNING HOMEWORK

Essential	1–5
Performance assessment	6–13
Portfolio	8, 10
Journal	10
Group	6, 7, 12
Review	14–19

MATERIALS

- graph paper, *optional*

▸ Helping with the Exercises

If you skipped the first part of Chapter 8 (parametric equations), do not assign Exercise 10d.

Exercise 1d The distance will be in terms of d.

Exercise 2 Students may be tempted to guess and check with integer values of y. Although this approach shows an understanding of the question, encourage use of the distance formula to express other values of y.

LESSON 9.1 Using the Distance Formula 491

ASSESSING PROGRESS indicates prerequisite skills students will use as they complete the investigation and understand the examples.

In **CLOSING THE LESSON,** the new mathematical ideas are reviewed and related.

BUILDING UNDERSTANDING suggests how teachers might assign exercises and how to anticipate difficulties and offer help.

ASSIGNING HOMEWORK CHART indicates ways to use the exercises.

ANNOTATIONS that appear in magenta on the student page or in wrap-around text give answers or suggestions for possible answers.

EXPLORATIONS present students with a guided activity followed by a few analysis or application questions.

Optional **TECHNOLOGY EXPLORATIONS** allow each teacher to customize the curriculum and give students the opportunity to use Sketchpad or Fathom to apply and extend their learning.

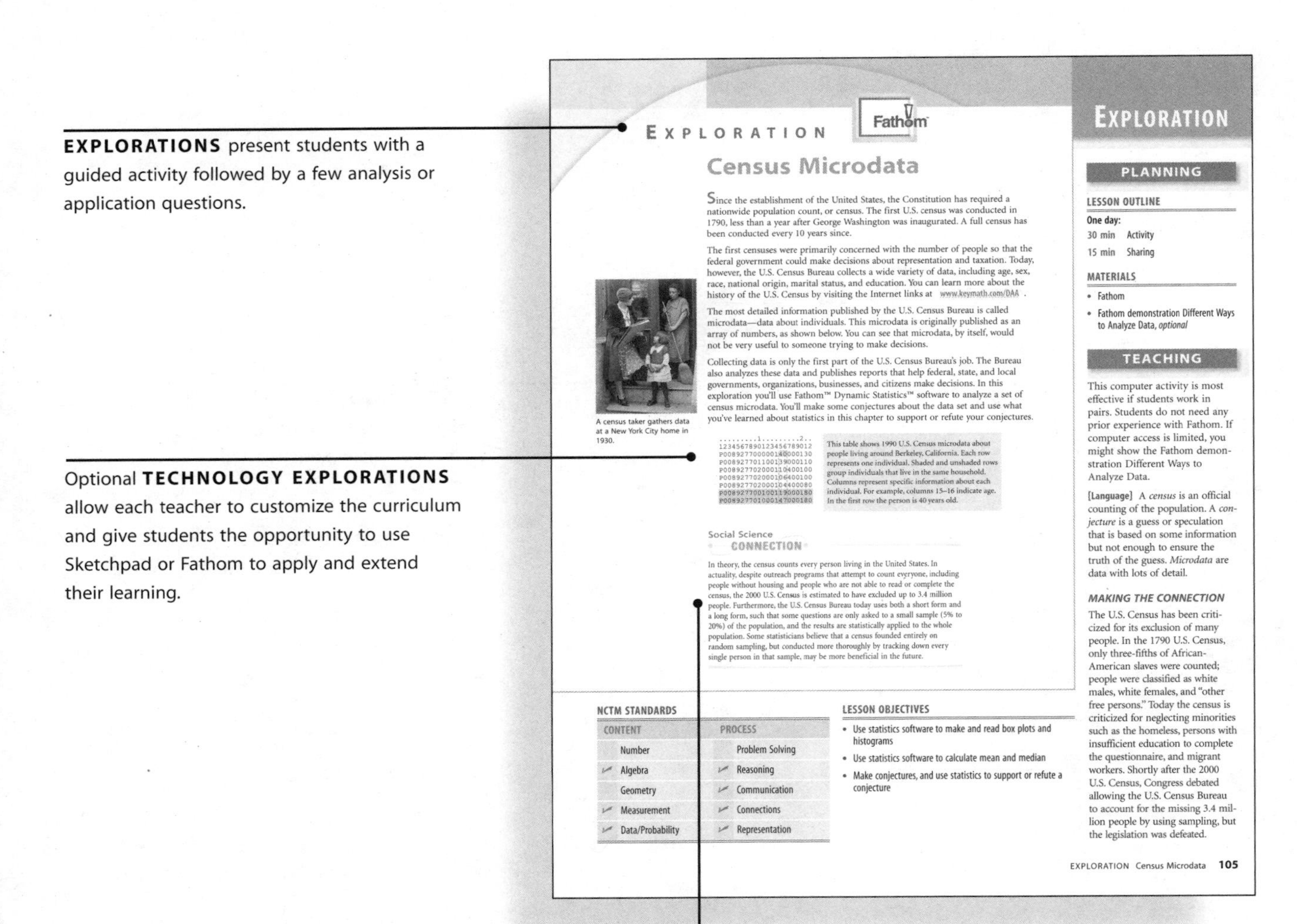

EXPLORATION Fathom™

Census Microdata

Since the establishment of the United States, the Constitution has required a nationwide population count, or census. The first U.S. census was conducted in 1790, less than a year after George Washington was inaugurated. A full census has been conducted every 10 years since.

The first censuses were primarily concerned with the number of people so that the federal government could make decisions about representation and taxation. Today, however, the U.S. Census Bureau collects a wide variety of data, including age, sex, race, national origin, marital status, and education. You can learn more about the history of the U.S. Census by visiting the Internet links at www.keymath.com/DAA .

The most detailed information published by the U.S. Census Bureau is called microdata—data about individuals. This microdata is originally published as an array of numbers, as shown below. You can see that microdata, by itself, would not be very useful to someone trying to make decisions.

Collecting data is only the first part of the U.S. Census Bureau's job. The Bureau also analyzes these data and publishes reports that help federal, state, and local governments, organizations, businesses, and citizens make decisions. In this exploration you'll use Fathom™ Dynamic Statistics™ software to analyze a set of census microdata. You'll make some conjectures about the data set and use what you've learned about statistics in this chapter to support or refute your conjectures.

A census taker gathers data at a New York City home in 1930.

```
.........1.........2..
1234567890123456789012
P00892770000014000013O
P00892770110013900011O
P00892770200011040010O
P00892770200010640010O
P00892770200010440008O
P00892770010011900018O
P00892770100014700018O
```

This table shows 1990 U.S. Census microdata about people living around Berkeley, California. Each row represents one individual. Shaded and unshaded rows group individuals that live in the same household. Columns represent specific information about each individual. For example, columns 15–16 indicate age. In the first row the person is 40 years old.

Social Science CONNECTION

In theory, the census counts every person living in the United States. In actuality, despite outreach programs that attempt to count everyone, including people without housing and people who are not able to read or complete the census, the 2000 U.S. Census is estimated to have excluded up to 3.4 million people. Furthermore, the U.S. Census Bureau today uses both a short form and a long form, such that some questions are only asked to a small sample (5% to 20%) of the population, and the results are statistically applied to the whole population. Some statisticians believe that a census founded entirely on random sampling, but conducted more thoroughly by tracking down every single person in that sample, may be more beneficial in the future.

NCTM STANDARDS

CONTENT	PROCESS
Number	Problem Solving
✓ Algebra	✓ Reasoning
Geometry	✓ Communication
✓ Measurement	✓ Connections
✓ Data/Probability	✓ Representation

LESSON OBJECTIVES

- Use statistics software to make and read box plots and histograms
- Use statistics software to calculate mean and median
- Make conjectures, and use statistics to support or refute a conjecture

EXPLORATION

PLANNING

LESSON OUTLINE

One day:

30 min Activity

15 min Sharing

MATERIALS

- Fathom
- Fathom demonstration Different Ways to Analyze Data, *optional*

TEACHING

This computer activity is most effective if students work in pairs. Students do not need any prior experience with Fathom. If computer access is limited, you might show the Fathom demonstration Different Ways to Analyze Data.

[Language] A *census* is an official counting of the population. A *conjecture* is a guess or speculation that is based on some information but not enough to ensure the truth of the guess. *Microdata* are data with lots of detail.

MAKING THE CONNECTION

The U.S. Census has been criticized for its exclusion of many people. In the 1790 U.S. Census, only three-fifths of African-American slaves were counted; people were classified as white males, white females, and "other free persons." Today the census is criticized for neglecting minorities such as the homeless, persons with insufficient education to complete the questionnaire, and migrant workers. Shortly after the 2000 U.S. Census, Congress debated allowing the U.S. Census Bureau to account for the missing 3.4 million people by using sampling, but the legislation was defeated.

EXPLORATION Census Microdata 105

CONNECTIONS extend students' appreciation of mathematics without diverting them from the essential material in the text. Some highlight how algebra or geometry applies to students' lives or future careers. Others give interesting facts from the history of mathematics.

PROJECTS often use Sketchpad or Fathom and attract students to the extended creative and practical possibilities of mathematics.

Additional information and answers are presented for **TAKE ANOTHER LOOK.**

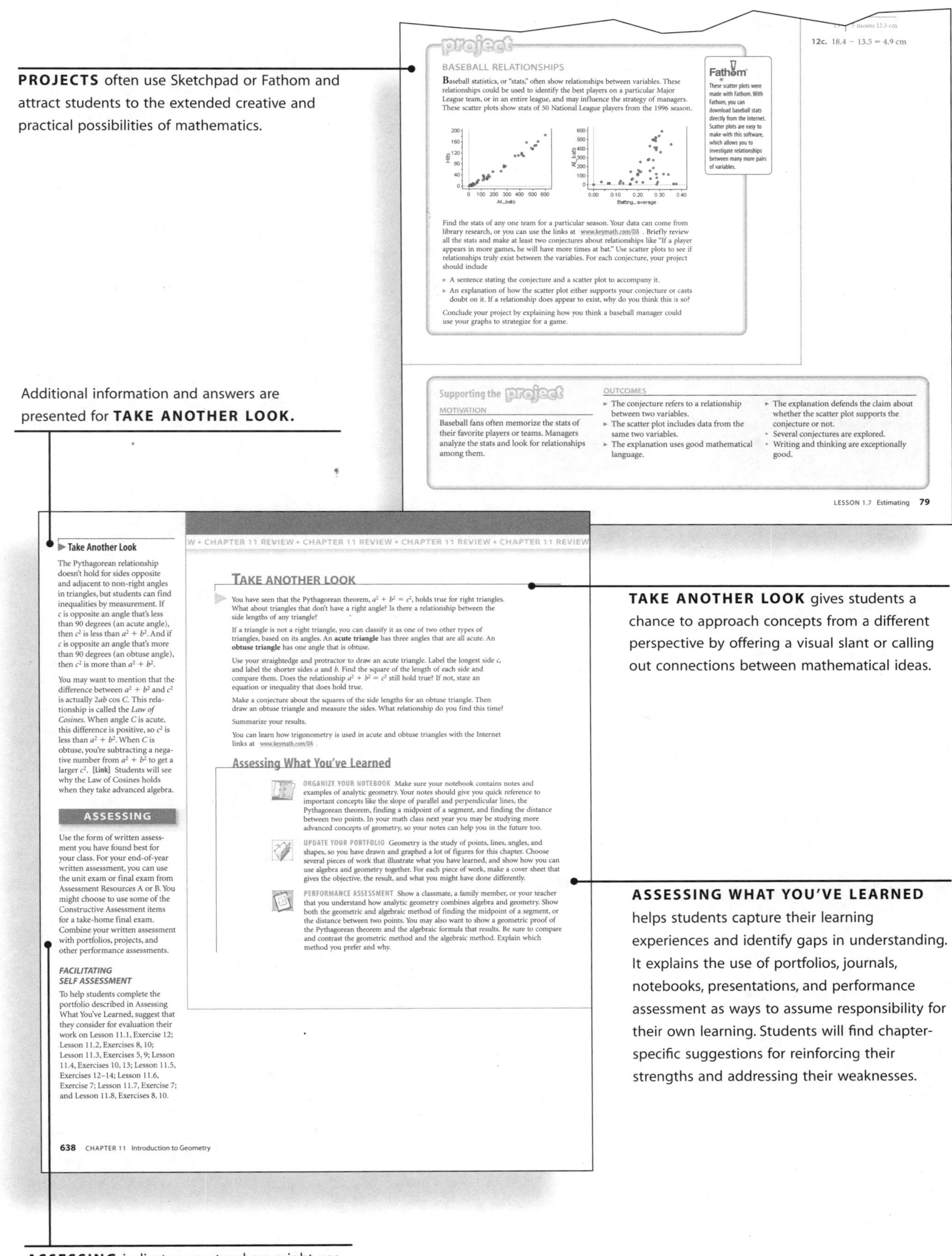

12c. $18.4 - 13.5 = 4.9$ cm

project

BASEBALL RELATIONSHIPS

Baseball statistics, or "stats," often show relationships between variables. These relationships could be used to identify the best players on a particular Major League team, or in an entire league, and may influence the strategy of managers. These scatter plots show stats of 50 National League players from the 1996 season.

Fathom

These scatter plots were made with Fathom. With Fathom, you can download baseball stats directly from the Internet. Scatter plots are easy to make with this software, which allows you to investigate relationships between many more pairs of variables.

Find the stats of any one team for a particular season. Your data can come from library research, or you can use the links at www.keymath.com/DA . Briefly review all the stats and make at least two conjectures about relationships like "If a player appears in more games, he will have more times at bat." Use scatter plots to see if relationships truly exist between the variables. For each conjecture, your project should include

- A sentence stating the conjecture and a scatter plot to accompany it.
- An explanation of how the scatter plot either supports your conjecture or casts doubt on it. If a relationship does appear to exist, why do you think this is so?

Conclude your project by explaining how you think a baseball manager could use your graphs to strategize for a game.

Supporting the project

MOTIVATION

Baseball fans often memorize the stats of their favorite players or teams. Managers analyze the stats and look for relationships among them.

OUTCOMES

- The conjecture refers to a relationship between two variables.
- The scatter plot includes data from the same two variables.
- The explanation uses good mathematical language.
- The explanation defends the claim about whether the scatter plot supports the conjecture or not.
- Several conjectures are explored.
- Writing and thinking are exceptionally good.

LESSON 1.7 Estimating 79

Take Another Look

The Pythagorean relationship doesn't hold for sides opposite and adjacent to non-right angles in triangles, but students can find inequalities by measurement. If c is opposite an angle that's less than 90 degrees (an acute angle), then c^2 is less than $a^2 + b^2$. And if c is opposite an angle that's more than 90 degrees (an obtuse angle), then c^2 is more than $a^2 + b^2$.

You may want to mention that the difference between $a^2 + b^2$ and c^2 is actually $2ab \cos C$. This relationship is called the *Law of Cosines*. When angle C is acute, this difference is positive, so c^2 is less than $a^2 + b^2$. When C is obtuse, you're subtracting a negative number from $a^2 + b^2$ to get a larger c^2. [Link] Students will see why the Law of Cosines holds when they take advanced algebra.

ASSESSING

Use the form of written assessment you have found best for your class. For your end-of-year written assessment, you can use the unit exam or final exam from Assessment Resources A or B. You might choose to use some of the Constructive Assessment items for a take-home final exam. Combine your written assessment with portfolios, projects, and other performance assessments.

FACILITATING SELF ASSESSMENT

To help students complete the portfolio described in Assessing What You've Learned, suggest that they consider for evaluation their work on Lesson 11.1, Exercise 12; Lesson 11.2, Exercises 8, 10; Lesson 11.3, Exercises 5, 9; Lesson 11.4, Exercises 10, 13; Lesson 11.5, Exercises 12–14; Lesson 11.6, Exercise 7; Lesson 11.7, Exercise 7; and Lesson 11.8, Exercises 8, 10.

W • CHAPTER 11 REVIEW • CHAPTER 11 REVIEW • CHAPTER 11 REVIEW • CHAPTER 11 REVIEW

TAKE ANOTHER LOOK

You have seen that the Pythagorean theorem, $a^2 + b^2 = c^2$, holds true for right triangles. What about triangles that don't have a right angle? Is there a relationship between the side lengths of any triangle?

If a triangle is not a right triangle, you can classify it as one of two other types of triangles, based on its angles. An **acute triangle** has three angles that are all acute. An **obtuse triangle** has one angle that is obtuse.

Use your straightedge and protractor to draw an acute triangle. Label the longest side c, and label the shorter sides a and b. Find the square of the length of each side and compare them. Does the relationship $a^2 + b^2 = c^2$ still hold true? If not, state an equation or inequality that does hold true.

Make a conjecture about the squares of the side lengths for an obtuse triangle. Then draw an obtuse triangle and measure the sides. What relationship do you find this time?

Summarize your results.

You can learn how trigonometry is used in acute and obtuse triangles with the Internet links at www.keymath.com/DA .

Assessing What You've Learned

ORGANIZE YOUR NOTEBOOK Make sure your notebook contains notes and examples of analytic geometry. Your notes should give you quick reference to important concepts like the slope of parallel and perpendicular lines, the Pythagorean theorem, finding a midpoint of a segment, and finding the distance between two points. In your math class next year you may be studying more advanced concepts of geometry, so your notes can help you in the future too.

UPDATE YOUR PORTFOLIO Geometry is the study of points, lines, angles, and shapes, so you have drawn and graphed a lot of figures for this chapter. Choose several pieces of work that illustrate what you have learned, and show how you can use algebra and geometry together. For each piece of work, make a cover sheet that gives the objective, the result, and what you might have done differently.

PERFORMANCE ASSESSMENT Show a classmate, a family member, or your teacher that you understand how analytic geometry combines algebra and geometry. Show both the geometric and algebraic method of finding the midpoint of a segment, or the distance between two points. You may also want to show a geometric proof of the Pythagorean theorem and the algebraic formula that results. Be sure to compare and contrast the geometric method and the algebraic method. Explain which method you prefer and why.

638 CHAPTER 11 Introduction to Geometry

TAKE ANOTHER LOOK gives students a chance to approach concepts from a different perspective by offering a visual slant or calling out connections between mathematical ideas.

ASSESSING WHAT YOU'VE LEARNED helps students capture their learning experiences and identify gaps in understanding. It explains the use of portfolios, journals, notebooks, presentations, and performance assessment as ways to assume responsibility for their own learning. Students will find chapter-specific suggestions for reinforcing their strengths and addressing their weaknesses.

ASSESSING indicates way teachers might use assessment resources and recommends way to help students with their self-assessment.

APPENDIX D Program Components

A comprehensive set of teaching resources supports each book in the *Discovering Mathematics* series. The following list briefly explains each resource. All resources are available for each book in the series, except as otherwise noted.

Assessment and Test Prep Resources

Assessment Resources A, Assessment Resources B

These books include forms A and B of quizzes (two or three per chapter), tests (one per chapter), constructive assessment options (three or four more in-depth problems per chapter), exams (one for every three chapters, corresponding to the student edition Mixed Reviews), and a final exam. All items except quiz items are included on the *Test Generator and Worksheet Builder* CD. The complete books are included on the *Teaching Resources* CD as pdf files.

Test Generator and Worksheet Builder CD

The CD contains a bank of hundreds of items, including all the *Assessment Resources* items except the quiz items. The software allows teachers to create their own worksheets, adding their own items or modifying existing items. It comes packaged with a *Quick Start Guide,* which has step-by-step instructions for getting started. For the states listed below, the *Test Generator* is aligned to the state standards. Teachers can search for items by the standard they want to practice. A practice test of about 20 items is included for these states: Arkansas, Florida, Illinois, Louisiana, and Oklahoma.

Discovering Mathematics Test Preparation (one book for series)

This booklet of 15 quizzes and 3 tests is designed to give practice for standardized test-taking, but it is correlated specifically to the *Discovering Mathematics* series, the National Assessment of Educational Progress (required by No Child Left Behind), and the TerraNova (the revised version of the California Achievement Test).

Discovering Mathematics Test Preparation: Florida Comprehensive Assessment Test (FCAT) (one book for series)

For Florida only, this booklet of 12 quizzes and 3 tests is correlated to both the FCAT and the *Discovering Mathematics* series.

Practice Resources

More Practice Your Skills with Answers (for *Discovering Algebra* and *Discovering Advanced Algebra*)

Practice Your Skills with Answers (for *Discovering Geometry*)

These books contain one page of exercises for each lesson. (Optional lessons may be combined or excluded.) The exercises practice basic skills and important procedures covered in the lesson, with and without context. The worksheets are available on **www.keymath.com** as downloadable pdf files, without answers.

More Practice Your Skills Student Workbook (for *Discovering Algebra* and *Discovering Advanced Algebra*)

Practice Your Skills Student Workbook (for *Discovering Geometry*)

These are consumable versions of *More Practice Your Skills with Answers* and *Practice Your Skills with Answers.* They do not contain answers.

Worksheet Builder CD (for *Discovering Algebra* only)

This software allows students or parents to create their own worksheets, adding their own items or modifying existing items. It comes packaged with a *Quick Start Guide,* which has step-by-step instructions for getting started. (Note: This CD is exactly like the *Test Generator* CD, except that it does not include our *Assessment Resources* items and is not aligned to state standards.)

Technology Resources

For *Discovering Algebra*

Calculator Notes for the Texas Instruments TI-73

Calculator Notes for the Texas Instruments TI-82

Calculator Notes for the Texas Instruments TI-83 and TI-83/84 Plus

Calculator Notes for the Casio FX 1.0 Plus and Algebra FX 2.0

Calculator Notes for the Casio FX-7400G Plus

Calculator Notes for the Casio CFX-9850GB Plus

These keystroke guides coach students in calculator language and mechanics. They are available at www.keypress.com/DM as downloadable pdf files.

Calculator Programs and Data CD

This CD includes data for student edition investigations, examples, and exercises, ready to load into each type of calculator, and programs to enhance student understanding of important topics. It is available at www.keypress.com/DM as downloadable calculator files.

For *Discovering Geometry*

Discovering Geometry with The Geometer's Sketchpad (includes CD with sketches)

This book contains *Discovering Geometry* lessons adapted for use with The Geometer's Sketchpad.

Demonstrations with The Geometer's Sketchpad (includes CD with sketches)

This book and CD demonstrates important concepts with The Geometer's Sketchpad and includes many investigations, examples, and exercises. It can be demonstrated by a teacher with an overhead, or used individually by students. It includes prepared sketches for student edition Sketchpad explorations.

For *Discovering Advanced Algebra*

Calculator Notes for the Texas Instruments TI-83 and TI-83/84 Plus

Calculator Notes for the Texas Instruments TI-89, TI-92, and Voyage 200

These keystroke guides coach students in calculator language and mechanics. They are available at www.keypress.com as downloadable pdf files. Calculator notes for the Casio CFX 9850 GB Plus are available at the same Web site.

Calculator Programs and Data CD

This CD includes data for student edition investigations, examples, and exercises, ready to load into each type of calculator, and programs to enhance student understanding of important topics. It is available at www.keypress.com/DM as downloadable calculator files.

Demonstrations with Fathom and The Geometer's Sketchpad (includes CD with sketches and Fathom files)

The Geometer's Sketchpad or Fathom Dynamic Statistics are used to demonstrate important concepts including those from explorations, examples, and exercises in the student edition. A teacher using an overhead projector to display a sketch or document from the CD can lead the class through a discussion, or students can explore these demonstrations alone or in groups.

Additional Teaching Resources

Solutions Manual

Students and teachers can use these worked-out solutions to all student edition exercises, exploration questions, Improving Your Skills puzzles, and Take Another Look activities.

Teaching and Worksheet Masters

This is a book of blackline masters of investigation worksheets, sample data sets, and some exercises, suitable for making transparencies to facilitate class discussion.

Condensed Lessons for Make-Up Work

For students who miss class, this book contains direct-teaching lessons that help students learn the important concepts in a lesson.

Condensed Lessons in Spanish

Spanish-speaking students will benefit from the direct-teaching lessons from all three books translated into Spanish.

Teaching Resources on CD

This CD contains all the printed teaching resources for each book as readable pdf files. *Discovering Geometry* and *Discovering Advanced Algebra* also include The Geometer's Sketchpad sketches, Fathom documents, and calculator programs and data.

Discovering Geometry: More Projects and Explorations

These additional projects and explorations supplement the geometry curriculum, including guided versions of some student edition projects. The Logic chapter and all the Cooperative Problem Solving activities and additional projects from *Discovering Geometry,* Second Edition, can be found at www.keypress.com/DM .

Implementation Resources

Discovering Mathematics: A Guide for Teachers

This helpful booklet gives teachers valuable information on implementing, using, and assessing student work for *Discovering Algebra, Discovering Geometry,* and *Discovering Advanced Algebra.*

APPENDIX E Research References

Association for Supervision and Curriculum Development. "Portfolios: Helping Students Think About Their Thinking." *Education Update.* Alexandria, VA: Association for Supervision and Curriculum Development, 2000.

Blakey, E., and S. Spence. "Developing Metacognition." *ERIC Digest.* http://www.ed.gov/databases/ERIC_Digests/ed327218.html, 1990.

Bransford, J. D., A. L. Brown, and R. R. Cocking. *How People Learn: Brain, Mind, Experience, and School.* Washington, D.C.: National Academy Press, 1999.

Caine, R. N., and G. Caine. *Unleashing the Power of Perceptual Change: The Potential of Brain-Based Teaching.* Alexandria, VA: Association for Supervision and Curriculum Development, 1997.

Clements, D. H., and M. T. Battista. "Geometry and Spatial Reasoning." In *Handbook of Research in Mathematics Teaching and Learning,* pp. 420–64. New York: NCTM/Macmillan, 1992.

de Villiers, Michael D. *Rethinking Proof with the Geometer's Sketchpad.* Emeryville, CA: Key Curriculum Press, 2003.

——. "*Review of Discovering Geometry* by Michael Serra, 2nd ed." Working paper for Key Curriculum Press, February 2000.

Donovan, M. Suzanne, John D. Bransford, and James W. Pellegrino, eds. *How People Learn: Bridging Research and Practice.* Washington, D.C.: National Academy Press, 1999.

Germain-McCarthy, Yvelyne. *Bringing the NCTM Standards to Life: Exemplary Practices from High Schools.* Larchmont, NY: Eye on Education, 1999.

Jensen, Eric. "How Julie's Brain Works." *Educational Leadership.* Alexandria, VA: Association for Supervision and Curriculum Development, November 1998.

National Council of Teachers of Mathematics. *Principles and Standards for School Mathematics.* Reston, VA: National Council of Teachers of Mathematics, 2000.

Resnick, L. B., and L. E. Klopfer, eds. *Toward the Thinking Curriculum: Current Cognitive Research.* Yearbook of the Association for Supervision and Curriculum Development. Alexandria, VA: Association for Supervision and Curriculum Development, 1989.

Westwater, Anne, and Pat Wolfe. "The Brain-Compatible Curriculum." *Educational Leadership.* Alexandria, VA: Association for Supervision and Curriculum Development, November 2000.

Wolfe, Pat. *Brain Matters: Translating Research into Classroom Practice.* Alexandria, VA: Association for Supervision and Curriculum Development, 2001.